**Springer Theses**

Recognizing Outstanding Ph.D. Research

## Aims and Scope

The series "Springer Theses" brings together a selection of the very best Ph.D. theses from around the world and across the physical sciences. Nominated and endorsed by two recognized specialists, each published volume has been selected for its scientific excellence and the high impact of its contents for the pertinent field of research. For greater accessibility to non-specialists, the published versions include an extended introduction, as well as a foreword by the student's supervisor explaining the special relevance of the work for the field. As a whole, the series will provide a valuable resource both for newcomers to the research fields described, and for other scientists seeking detailed background information on special questions. Finally, it provides an accredited documentation of the valuable contributions made by today's younger generation of scientists.

## Theses are accepted into the series by invited nomination only and must fulfill all of the following criteria

- They must be written in good English.
- The topic should fall within the confines of Chemistry, Physics, Earth Sciences, Engineering and related interdisciplinary fields such as Materials, Nanoscience, Chemical Engineering, Complex Systems and Biophysics.
- The work reported in the thesis must represent a significant scientific advance.
- If the thesis includes previously published material, permission to reproduce this must be gained from the respective copyright holder.
- They must have been examined and passed during the 12 months prior to nomination.
- Each thesis should include a foreword by the supervisor outlining the significance of its content.
- The theses should have a clearly defined structure including an introduction accessible to scientists not expert in that particular field.

Indexed by zbMATH.

More information about this series at http://www.springer.com/series/8790

Masato Hamada

# Theory of Generation and Conversion of Phonon Angular Momentum

Doctoral Thesis accepted by
Tokyo Institute of Technology, Tokyo, Japan

Springer

*Author*
Dr. Masato Hamada
Department of Physics
Tokyo Institute of Technology
Tokyo, Japan

*Supervisor*
Prof. Shuichi Murakami
Department of Physics
Tokyo Institute of Technology
Tokyo, Japan

ISSN 2190-5053 ISSN 2190-5061 (electronic)
Springer Theses
ISBN 978-981-33-4689-5 ISBN 978-981-33-4690-1 (eBook)
https://doi.org/10.1007/978-981-33-4690-1



This Springer imprint is published by the registered company Springer Nature Singapore Pte Ltd.
The registered company address is: 152 Beach Road, #21-01/04 Gateway East, Singapore 189721, Singapore

# Supervisor's Foreword

Phonons are vibrations of atoms constituting a crystal, and they propagate inside the crystal like waves. Their properties have been well studied experimentally and theoretically in various materials. While they are usually regarded as vibrations, they can also have rotational motions, representing elliptic cyclic motions of atoms, as has been intensively studied recently. One can also associate phonon angular momenta to these rotational motions. Whether or not a crystal can have phonon angular momentum depends on symmetries of the crystal, and therefore can possibly be controlled by external fields. Moreover, since phonons are rotational motions similar to electron spins, one can expect coupling between them from the symmetry viewpoint. Such ideas will enable us to control and utilize phonon angular momenta as a new degree of freedom in crystals.

This book is concerned with theoretical investigations of fundamental properties of phonon angular momenta in crystals. This book, based on the dissertation of Dr. Masato Hamada, theoretically proposes new ways of generation of angular momenta of phonons by external fields and of conversion from phonon angular momentum into electron spins. By combining microscopic calculation, symmetry analysis, and first-principle calculations, it is predicted in this book that heat current and electric field will generate phonon angular momenta in crystals depending on symmetries of the crystal. It is also shown that electron spin magnetization is dynamically generated from rotational motions of atoms in the phonons. The theory here is widely applicable to a wide class of materials, and therefore the theoretical predictions here can stimulate future theoretical and experimental studies on phonons.

Tokyo, Japan
June 2020

Shuichi Murakami

**Parts of this thesis have been published in the following journal articles:**

1. Masato Hamada, Takehito Yokoyama, and Shuichi Murakami, "Spin current generation and magnetic response in carbon nanotubes by the twiitng phonon mode", Phys. Rev. B **92**, 060409 (2015).
   DOI: https://doi.org/10.1103/PhysRevB.92.060409.
   Copyright © 2015 American Physical Society.
2. Masato Hamada, Emi Minamitani, Motoaki Hirayama, and Shuichi Murakami, "Phonon Angular Momentum Induced by the Temperature Gradient", Phys. Rev. Lett. **121**, 175301 (2018).
   DOI: https://doi.org/10.1103/PhysRevLett.121.175301.
   Copyright © 2018 American Physical Society.
3. Masato Hamada and Shuichi Murakami, "Phonon rotoelectric effect", Phys. Rev. B **101**, 144306 (2020).
   DOI: https://doi.org/10.1103/PhysRevB.101.144306.
   Copyright © 2020 American Physical Society.
4. Masato Hamada and Shuichi Murakami, "Conversion between electron spin and microscopic atomic rotation", Phys. Rev. Research. **2**, 023275 (2020).
   DOI: https://doi.org/10.1103/PhysRevResearch.2.023275.
   Copyright © 2020 American Physical Society.

# Acknowledgements

First, I would like to express the deepest appreciation to Prof. Shuichi Murakami for his helpful advice, incisive comments, fruitful discussion, and thoughtful guidance over the last 6 years. When I was faced with difficult problems in my study and I am troubled with the development of my study, he found time among his tight schedule for discussions, shared problems, and suggested me incisive advices. I have learned from him methods to approach the research, the way of presentation, and methods to write scientific papers. Furthermore, I could attend various seminars and conferences and interact with many researchers.

I would like to thank a lot Dr. Minamitani Emi for her thoughtful guidance, constructive comments, and fruitful discussions. Although it was a short period, I have learned from her how to do the first-principle calculation with her kind guidance and appropriate advice. Through the collaboration with her, I have developed our study with fruitful discussions. I would like to thank a lot Dr. Motoaki Hirayama also whose comments made enormous contributions to my study. I would like to thank all the members of the Murakami Group for fruitful discussion.

Special thanks are due to Prof. Eiji Saitoh and his group members, and to Prof. Keiji Saito for helpful discussions and useful suggestion. I have greatly benefited from their valuable advice and comments.

Finally, I would like to thank my parents and brothers for always encouraging me.

# Contents

**1 Introduction** . . . . . 1
1.1 Introduction . . . . . 1
1.2 Purpose of This Thesis . . . . . 2
1.3 Organization of This Thesis . . . . . 3
References . . . . . 4

**2 Background** . . . . . 7
2.1 Phonon Angular Momentum . . . . . 7
2.1.1 Overview of Lattice Dynamics and Phonons in the Harmonic Approximation . . . . . 7
2.1.2 Formalism of the Phonon Angular Momentum . . . . . 11
2.1.3 Time-Reversal and Inversion Symmetries . . . . . 13
2.1.4 Phonon System Without Time-Reversal Symmetry . . . . . 13
2.1.5 Phonon Angular Momentum in the High- and Low-Temperature Limits . . . . . 18
2.1.6 Chiral Phonons in Monolayer Hexagonal Lattices . . . . . 19
2.2 Cross-Correlated Response . . . . . 22
2.2.1 Edelstein Effect . . . . . 25
2.2.2 Magnetoelectric Effect . . . . . 26
References . . . . . 26

**3 Phonon Thermal Edelstein Effect** . . . . . 29
3.1 Phonon Thermal Edelstein Effect . . . . . 29
3.2 Symmetry Constraint for the Phonon Edelstein Effect . . . . . 31
3.3 Generation of Phonon Angular Momentum by Temperature Gradient . . . . . 33
3.3.1 Polar Crystal: Wurtzite Gallium Nitride . . . . . 33
3.3.2 Chiral Crystals: Tellurium and Selenium . . . . . 38

3.4 Conversion of Phonon Angular Momentum Generated by Temperature Gradient . . . . . 41
3.4.1 Conversion to a Rigid-Body Rotation of Crystal . . . . . 41
3.4.2 Conversion to a Magnetization . . . . . 43
3.5 Phonon Thermal Edelstein Effect in the High- and Low-Temperature Limits . . . . . 44
3.6 Summary . . . . . 47
References . . . . . 47

**4 Magnetoelectric Effect for Phonons** . . . . . 49
4.1 Phonon Angular Momentum with Neither Time-Reversal nor Inversion Symmetries . . . . . 49
4.2 Formalism of Phonons with the Spin-Phonon Interaction . . . . . 50
4.3 Magnetoelectric Effect for Phonons . . . . . 54
4.3.1 Form of the Response Tensor Using the Magnetic Point Group in This Toy Model . . . . . 57
4.3.2 Numerical Results . . . . . 59
4.4 Phonon Angular Momentum Due to the Magnetoelectric Effect in the High- and the Low-Temperature Limits . . . . . 67
4.5 Summary . . . . . 68
References . . . . . 68

**5 Conversion Between Spins and Mechanical Rotations** . . . . . 69
5.1 Conversion Between Spins and Mechanical Rotations . . . . . 69
5.2 Honeycomb-Lattice Model with the Periodic Microscopic Local Rotation . . . . . 69
5.3 Adiabatic Series Expansion . . . . . 73
5.4 Expectation Values in the Adiabatic Approximation . . . . . 77
5.5 Expectation Values of Spin Operator in Our Model . . . . . 80
5.6 Discussion . . . . . 83
5.7 Summary . . . . . 87
References . . . . . 87

**6 Conclusion** . . . . . 89

**Curriculum Vitae** . . . . . 93

# Chapter 1
# Introduction

## 1.1 Introduction

Various excitations in solids have been among the most important subjects in solid state physics, because they exhibit various novel properties in a variety of materials. Representative examples are electrons, photons, phonons, and magnons. In quantum theory, light and electromagnetic waves have both wave nature and particle nature, and it is called photon. Similarly, the lattice vibration of each atom in crystals and the propagating waves have both particle and wave natures, and it is called phonon. Here, phonons represent collective vibrations, and they are not real particles. Unlike photons, phonons are classified into longitudinal and transverse waves. Phonons are classified into acoustic modes, where the neighboring atoms vibrate in the same phases, and optical modes, where the neighboring atoms vibrate in opposite phase, when there are two or more atoms in a unit cell. Phonons have been studied for a long time. Since phonons carry heat in solids, thermal conductivity and specific heat are closely related to a phonon dispersion in solids. It is also known that phonons interact with other elementary excitations. As an example of the interaction between electrons and phonons, in the jellium model, consisting of free electrons and uniform distribution of positive ions, a compression wave of a positive charge is generated by compression of waves in the longitudinal phonons. Then, the resulting periodic electric field having the same period with phonons combines with free electrons and attenuates sound waves. As another example, the electron-phonon interaction in a metal causes an effective attractive force between electrons, and it gives rise to superconductivity. For the interaction between photons and phonons, when a monochromatic light $\omega_0$ is injected into a solid, the photoelectric field produces the dipole moment in the solids $\boldsymbol{\mu} = \alpha \mathbf{E}_0 \cos \omega_0 t$ with the polarizability tensor $\alpha$. The phonons slightly alter the polarizability tensor at the phonon frequency $\omega$. Then, the dipole moment is modulated and has a vibration component of $\omega_0 \pm \omega$, and the light emitted therefrom has the same vibration component. This scattering phenomenon is called Raman scattering. As described above, interaction of phonons with other elementary excitations have been intensively studied.



M. Hamada, *Theory of Generation and Conversion of Phonon Angular Momentum*, Springer Theses, https://doi.org/10.1007/978-981-33-4690-1_1

In recent years, the control of phonon degree of freedom has been an active field for experimental and theoretical studies. Phononics is a new active field aiming at controlling phonon dynamics and information transport by phonons. One of the recent topics in phononics is the phonon Hall effect [1–9]. Apart from phonons, the Hall effect of the electrons system has been known for a long time. When a magnetic field is applied perpendicular to the electric current, the electromotive force appears in a direction orthogonal to both the electric current and the magnetic field due to the Lorentz force. In the phonon Hall effect, when a heat current is used instead of an electric current, a temperature gradient occurs in a direction perpendicular to both the heat current and the magnetic field. This phenomenon was experimentally observed in ionic paramagnetic dielectrics, such as $Ta_3Ga_5O_{12}$(TGG) [1, 2], and in a quantum spin liquid candidate $Ba_3CuSb_2O_9$ [3]. What is surprising is that, unlike electrons, phonons are quasiparticles with neither charges nor spins, and it cannot be directly coupled with external fields via Lorentz force. In one of the theoretical interpretations [6, 7], it is attributed to the spin-phonon interaction [10–14] which is the coupling between localized spin and lattice vibration. However, the experimental result and the theory is not consistent, and the fundamental mechanism is still not clear. Furthermore, new properties of phonons are proposed, leading to novel topological phenomena, analogous to quantum (anomalous) Hall effect and a topological insulator in electronics systems. The quantum (anomalous) Hall effect is characterized by a topological invariant called the Chern number of the first class, which is defined by an integral of the Berry curvature over the Brillouin zone and can be nonzero only for systems with broken time-reversal symmetry. The phononic states with nonzero Chern number give the quantum (anomalous) Hall-like edge states of phonons, which are topologically protected to be gapless, one-way, and immune to backscattering. Moreover, one of the merits in phononics systems is that topological modes can be realized in macroscopic systems, such as the mechanical honeycomb lattice [15–20], which consists of the springs and masses, and the sonic crystal [21, 22], which consists of triangular polymethyl methacrylate rods positioned in a triangular lattice. For example, it was theoretically proposed that the Chern number has nonzero values in the phonon system where the time-reversal symmetry is broken by a mechanical rotation [20].

With the discovery of the phonon Hall effect, the angular momentum of phonon was proposed [23–25]. Unlike the general macroscopic rigid-body rotation, the phonon angular momentum is understood as microscopic local rotations of the atoms around their equilibrium positions in crystals. By extending this idea, the concepts of chiral phonon [26], namely, the phonon with the angular momentum was theoretically proposed. Subsequently, the chiral phonon is observed in tungsten diselenide [27].

## 1.2 Purpose of This Thesis

In this thesis, we focus on how to generate this phonon angular momentum and to convert it to other degree of freedom. We focus on the conversion between the electron spin and the mechanical rotation. In magnets, the Einstein-de Haas effect [28] and the

Barnett effect [29] are known to convert between a magnetization and mechanical rotation. In the Einstein-de Haas effect when a magnetic material is magnetized by the external magnetic field, the magnetic material rotates around the external magnetic field to conserve the total angular momentum. The Barnett effect is a reciprocal effect to the Einstein-de Haas effect. The origin of spin conversion in these effect is considered to be the spin-rotation coupling [30–33] which couples spin and mechanical rotation. Moreover, in spintronics, a method to generate a spin current via spin-rotation coupling has been reported theoretically and experimentally, such as a surface acoustic wave [34, 35], the twisting vibrational mode in a carbon nanotube [36], and the flow of a liquid metal [37]. In this way, the conversion between a spin and a mechanical rotation has been studied intensively. We focus on the phonon angular momentum. Our first purpose is to investigate the microscopic mechanism of the conversion between the electron spin and the microscopic local rotation. We also investigate generation methods of phonon angular momentum, which is not well-known. Our second purpose is to clarify how to generate and to observe the phonon angular momentum and observation method. In the system without the inversion (time-reversal) symmetry, the phonon angular momentum of each phonon mode is an odd (even) function of wave vectors. In order to make the phonon angular momentum nonzero, the time-reversal or the inversion symmetries should be broken. In previous study [23], in the system without the time-reversal symmetry, such as the system under a magnetic field, the total phonon angular momentum have a finite value in equilibrium. On the other hand, in the system without the inversion symmetry, the total phonon angular momentum vanish because the phonon angular momentum of each phonon mode cancels between wave vectors $\mathbf{k}$ and $-\mathbf{k}$. In this thesis, we propose a new method to generate the phonon angular momentum in systems without the inversion symmetry. Moreover, since the phonon angular momentum cannot be directly observed, we propose conversion of the phonon angular momentum to other degree of freedom.

## 1.3 Organization of This Thesis

In this thesis, we discuss generation of the phonon angular momentum in non-magnetic crystals and magnetic crystals, and we propose experimental observation methods of the phonon angular momentum. Moreover, we discuss the spin magnetization due to the microscopic local rotation using a simple toy model. In Chap. 2, we introduce the background of the present work: an overview of the lattice dynamics and the phonon, a review of the phonon angular momentum, and a overview of cross-correlated response and symmetry. In Chaps. 3, 4, and 5, we develop our study. In Chap. 3, we discuss the generation of the phonon angular momentum by temperature gradient in the non-magnetic crystals without inversion symmetry [38]. We also show numerical results of the phonon angular angular momentum generated by temperature gradient in the wurtzite GaN, Te, and Se. Moreover, we propose experimental observation methods of the phonon angular momentum, such as the conversions to a rigid-body rotation in solids and to a magnetization. In Chap. 4, we discuss phonon angular momentum in the magnetic crystals without both the time-reversal and the inversion symmetries. We show the phonon angular momentum generated by the electric field in our toy model [39]. In Chap. 5, we discuss the microscopic mechanism of the conversion between the electron spin and the microscopic local rotation. We show that the electron spin expectation value along the microscopic rotational axis is proportional to the angular velocity of the microscopic local rotation in the adiabatic approximation [40]. In Chap. 6, we summarize this thesis.

## References


1. Strohm C, Rikken GLJA, Wyder P (2005) Phenomenological evidence for the phonon hall effect. Phys Rev Lett 95:155901
2. Inyushkin AV, Taldenkov AN (2007) On the phonon hall effect in a paramagnetic dielectric. JETP Lett 86:379–382
3. Sugii K et al (2017) Thermal hall effect in a phonon-glass $ba_3cusb_2o_9$. Phys Rev Lett 118:145902
4. Mori M, Spencer-Smith A, Sushkov OP, Maekawa S (2014) Origin of the phonon hall effect in rare-earth garnets. Phys Rev Lett 113:265901
5. Qin T, Zhou J, Shi J (2012) Berry curvature and the phonon hall effect. Phys Rev B 86:104305
6. Wang J-S, Zhang L (2009) Phonon hall thermal conductivity from the green-kubo formula. Phys Rev B 80:012301
7. Sheng L, Sheng DN, Ting CS (2006) Theory of the phonon hall effect in paramagnetic dielectrics. Phys Rev Lett 96:155901
8. Zhang L, Ren J, Wang J-S, Li B (2010) Topological nature of the phonon hall effect. Phys Rev Lett 105:225901
9. Agarwalla BK, Zhang L, Wang J-S, Li B (2011) Phonon hall effect in ionic crystals in the presence of static magnetic field. Eur Phys J B 81:197–202
10. de L Kronig R (1939) On the mechanism of paramagnetic relaxation. Physica 6:33–43
11. Van Vleck JH (1940) Paramagnetic relaxation times for titanium and chrome alum. Phys Rev 57:426–447
12. Capellmann H, Lipinski S (1991) Spin-phonon coupling in intermediate valency: exactly solvable models. Zeitschrift für Physik B Condens Matter 83:199–205
13. Capellmann H, Neumann KU (1987) Intermediate valency i: couplings, hamiltonian, and electrical resistivity. Zeitschrift für Physik B Condens Matter 67:53–61
14. Capellmann H, Lipinski S, Neumann K-U (1989) A microscopic model for the coupling of spin fluctuations and charge fluctuation in intermediate valency. Zeitschrift fur Physik B Condens Matter 75:323–329
15. Kariyado T, Hatsugai Y (2015) Manipulation of dirac cones in mechanical graphene. Sci Rep 5
16. Wang Y-T, Zhang S (2016) Elastic spin-hall effect in mechanical graphene. New J Phys 18:113014
17. Socolar JES, Kane CL, Lubensky TC (2017) Mechanical graphene. New J Phys 19:025003
18. Wang P, Lu L, Bertoldi K (2015) Topological phononic crystals with one-way elastic edge waves. Phys Rev Lett 115:104302
19. Nash LM et al (2015) Topological mechanics of gyroscopic metamaterials. Proc Natl Acad Sci 112:14495–14500
20. Wang Y-T, Luan P-G, Zhang S (2015) Coriolis force induced topological order for classical mechanical vibrations. New J Phys 17:073031

21. Lu J, Qiu C, Ke M, Liu Z (2016) Valley vortex states in sonic crystals. Phys Rev Lett 116:093901
22. Lu J et al (2017) Observation of topological valley transport of sound in sonic crystals. Nat Phys 13:369–374
23. Zhang L, Niu Q (2014) Angular momentum of phonons and the einstein-de haas effect. Phys Rev Lett 112:085503
24. Garanin DA, Chudnovsky EM (2015) Angular momentum in spin-phonon processes. Phys Rev B 92:024421
25. Zhang L (2016) Berry curvature and various thermal hall effects. New J Phys 18:103039
26. Zhang L, Niu Q (2015) Chiral phonons at high-symmetry points in monolayer hexagonal lattices. Phys Rev Lett 115:115502
27. Zhu H et al (2018) Observation of chiral phonons. Science 359:579–582
28. Einstein A, de Haas WJ (1915) Experimenteller nachweis der ampereschen molekularstrome 17:152–170
29. Barnett SJ (1915) Magnetization by rotation. Phys Rev 6:239–270
30. Matsuo M, Ieda J, Saitoh E, Maekawa S (2011) Effects of mechanical rotation on spin currents. Phys Rev Lett 106:076601
31. Matsuo M, Ieda J, Saitoh E, Maekawa S (2011) Spin-dependent inertial force and spin current in accelerating systems. Phys Rev B 84:104410
32. de Oliveira C, Tiomno J (1962) Representations of dirac equation in general relativity. Il Nuovo Cimento 24:672–687
33. Mashhoon B (1988) Neutron interferometry in a rotating frame of reference. Phys Rev Lett 61:2639–2642
34. Matsuo M, Ieda J, Harii K, Saitoh E, Maekawa S (2013) Mechanical generation of spin current by spin-rotation coupling. Phys Rev B 87:180402
35. Kobayashi D et al (2017) Spin current generation using a surface acoustic wave generated via spin-rotation coupling. Phys Rev Lett 119:077202
36. Hamada M, Yokoyama T, Murakami S (2015) Spin current generation and magnetic response in carbon nanotubes by the twisting phonon mode. Phys Rev B 92:060409
37. Takahashi R et al (2016) Spin hydrodynamic generation. Nat Phys 12:52
38. Hamada M, Minamitani E, Hirayama M, Murakami S (2018) Phonon angular momentum induced by the temperature gradient. Phys Rev Lett 121:175301
39. Hamada M, Murakami S (2020) Phonon rotoelectric effect. Phys Rev B 101:144306
40. Hamada M, Murakami S (2020) Conversion between electron spin and microscopic atomic rotation. Phys Rev Res 2:023275

# Chapter 2
# Background

## 2.1 Phonon Angular Momentum

In this section, we introduce the formalism of phonon angular momentum and its properties in systems without time-reversal symmetry [1]. We also explain the chirality of phonons at valley points in momentum space and show how to excite the chiral phonons [2], together with related results in experiments [3].

### *2.1.1 Overview of Lattice Dynamics and Phonons in the Harmonic Approximation*

In this section, for the purpose of understanding the phonon angular momentum, we give an overview of lattice dynamics and phonons in a harmonic approximation [4–7]. First, we introduce the dynamics of a crystal lattice. In Fig. 2.1, we show the schematic figure of a crystal lattice. We consider a crystal with $N$ atoms per unit cell, and $\mathbf{u}_{l\kappa}$ is the displacement vector of the $\kappa$th atom in the $l$th unit cell. In addition, we assume that the total potential energy $U$ of the crystal lattice is a function of the instantaneous position vectors $\mathbf{r}_{l\kappa}$ of all atoms. Here, the instantaneous position vector is given by $\mathbf{r}_{l\kappa} = \mathbf{R}_{l\kappa} + \mathbf{u}_{l\kappa}$, where $\mathbf{R}_{l\kappa}$ is an equilibrium position vector of the $\kappa$th atom in the $l$th unit cell. The total potential energy can be expanded in a Taylor series in powers of the small displacement around the equilibrium position:

$$
\begin{aligned}
U = U_0 &+ \sum_{l\kappa} \sum_{\alpha=x,y,z} \left.\frac{\partial U}{\partial u_{l\kappa,\alpha}}\right|_0 u_{l\kappa,\alpha} \\
&+ \frac{1}{2} \sum_{l\kappa} \sum_{l'\kappa'} \sum_{\alpha,\beta=x,y,z} \left.\frac{\partial^2 U}{\partial u_{l\kappa,\alpha} \partial u_{l'\kappa',\beta}}\right|_0 u_{l\kappa,\alpha} u_{l'\kappa',\beta} + \dots,
\end{aligned}
\tag{2.1}
$$


M. Hamada, *Theory of Generation and Conversion of Phonon Angular Momentum*, Springer Theses, https://doi.org/10.1007/978-981-33-4690-1_2

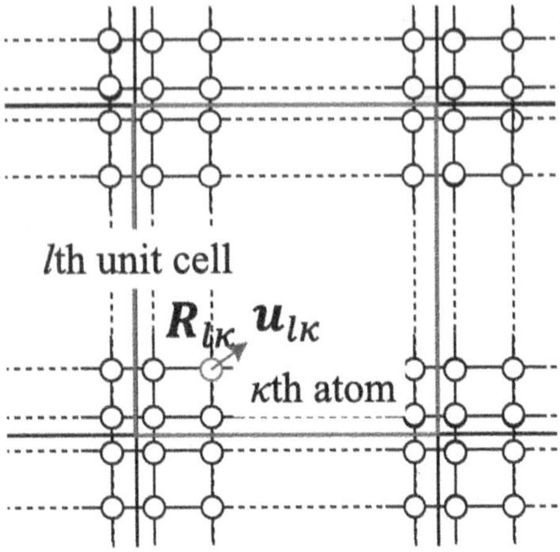


**Fig. 2.1** Schematic figure of a crystal lattice

where $|_0$ means that the derivative is evaluated for the equilibrium positions of the atoms. The first term is a constant. This is unimportant for lattice dynamics problems and can be set zero. The second term gives rise to a force and must vanish for the equilibrium position. The third term and other subsequent terms affect the lattice dynamics, and an approximation of lattice dynamics with retaining up to the quadratic terms is known as the harmonic approximation.

In the harmonic approximation, the potential energy is obtained as

$$U_{\text{harm}} = \frac{1}{2} \sum_{l\kappa} \sum_{l'\kappa'} \sum_{\alpha,\beta=x,y,z} \Phi_{\alpha\beta}(l\kappa, l'\kappa') u_{l\kappa,\alpha} u_{l'\kappa',\beta}, \tag{2.2}$$

$$\Phi_{\alpha\beta}(l\kappa, l'\kappa') = \left. \frac{\partial^2 U}{\partial u_{l\kappa,\alpha} \partial u_{l'\kappa',\beta}} \right|_0, \tag{2.3}$$

where $\Phi$ is called a force constant matrix. Thus, the Lagrangian of phonons in the harmonic approximation is represented as

$$L = \sum_{l\kappa} \left( \frac{1}{2} m_\kappa \dot{\mathbf{u}}_{l\kappa}^2 - \frac{1}{2} \sum_{l'\kappa'} \sum_{\alpha,\beta=x,y,z} \Phi_{\alpha\beta}(l\kappa, l'\kappa') u_{l\kappa,\alpha} u_{l'\kappa',\beta} \right), \tag{2.4}$$

and the equation of motion is written as

$$m_\kappa \ddot{u}_{l\kappa,\alpha} = -\sum_{l'\kappa',\beta} \Phi_{\alpha\beta}(l\kappa, l'\kappa') u_{l'\kappa',\beta}, \tag{2.5}$$

where $m_\kappa$ is the mass of the $\kappa$th atom in the unit cell.

The force constant matrix $\Phi$ obeys two important symmetry relations. From the lattice translational symmetry, the force constant matrix satisfies

$$\Phi_{\alpha\beta}(l\kappa, l'\kappa') = \Phi_{\alpha\beta}(0\kappa, (l'-l)\kappa'). \tag{2.6}$$

From the infinitesimal translational invariance of the whole crystal, the force constant matrix satisfies

$$\Phi_{\alpha\beta}(l\kappa, l\kappa) = -\sum_{l'\kappa' \neq l\kappa} \Phi_{\alpha\beta}(l\kappa, l'\kappa'), \tag{2.7}$$

and this relation is called acoustic sum rule. To solve the equation of motion Eq. (2.5), we introduce the Fourier transformation of the displacement vector:

$$\sqrt{m_\kappa}\mathbf{u}_{l\kappa} = \frac{1}{\sqrt{N}}\sum_{\mathbf{k}} \mathbf{U}_{\mathbf{k}\kappa} e^{\mathrm{i}\mathbf{R}_l\cdot\mathbf{k}-\mathrm{i}\omega t}, \tag{2.8}$$

and Eq. (2.5) is rewritten as

$$\omega^2 U_{\mathbf{k}\kappa,\alpha} = \sum_{\kappa',\beta} D_{\alpha\beta}(\mathbf{k}, \kappa\kappa') U_{\mathbf{k}\kappa',\beta}, \tag{2.9}$$

$$D_{\alpha\beta}(\mathbf{k}, \kappa\kappa') = \sum_{l'} \frac{\Phi_{\alpha\beta}(l\kappa, l'\kappa')}{\sqrt{m_\kappa m_{\kappa'}}} e^{\mathrm{i}\mathbf{k}\cdot(\mathbf{R}_{l'}-\mathbf{R}_l)}, \tag{2.10}$$

where $D$ is called the dynamical matrix. From the definition of dynamical matrix Eq. (2.10) and that of force constant matrix $\Phi_{\alpha\beta}(l\kappa, l'\kappa')$, one has

$$D^*_{\alpha\beta}(\mathbf{k}, \kappa\kappa') = D_{\beta\alpha}(\mathbf{k}, \kappa'\kappa). \tag{2.11}$$

We can also get for wavevector $-\mathbf{k}$ from the definition of dynamical matrix Eq. (2.10)

$$D^*_{\alpha\beta}(\mathbf{k}, \kappa\kappa') = D_{\alpha\beta}(-\mathbf{k}, \kappa\kappa'). \tag{2.12}$$

Thus, the dynamical matrix is a $3N \times 3N$ Hermitian matrix. Equation (2.9) produces $3N$ eigenvalues $\omega^2_{\mathbf{k}\sigma}$, where $\sigma = 1, 2, \ldots, 3N$. Because the dynamical matrix $D$ is Hermitian, its eigenvalues $\omega^2_{\mathbf{k}\sigma}$ are real and we consider $\omega_{\mathbf{k}\sigma}$ to be real to guarantee the stability of the crystal.

We introduce the displacement polarization vector $\boldsymbol{\epsilon}_{\mathbf{k}\sigma\kappa}$ which is an eigenvector for the eigenvalue $\omega^2_{\mathbf{k}\sigma}$. Therefore, Eq. (2.9) can be rewritten as

$$\omega^2_{\mathbf{k}\sigma}\epsilon_{\mathbf{k}\sigma\kappa,\alpha} = \sum_{\kappa'\beta} D_{\alpha\beta}(\mathbf{k}, \kappa\kappa')\epsilon_{\mathbf{k}\sigma\kappa',\beta}. \tag{2.13}$$

Now we write the displacement polarization vector within the unit cell and the dynamical matrix within the unit cell:

$$\epsilon_{\mathbf{k}\sigma} = (\epsilon_{\mathbf{k}\sigma 1,x}, \epsilon_{\mathbf{k}\sigma 1,y}, \epsilon_{\mathbf{k}\sigma 1,z}, \ldots, \epsilon_{\mathbf{k}\sigma N,z,})^T, \tag{2.14}$$

$$D(\mathbf{k}) = \begin{pmatrix} D(\mathbf{k}, 11) & D(\mathbf{k}, 12) & \ldots & D(\mathbf{k}, 1N) \\ D(\mathbf{k}, 21) & D(\mathbf{k}, 22) & \ldots & D(\mathbf{k}, 2N) \\ \vdots & \vdots & \ddots & \vdots \\ D(\mathbf{k}, N1) & D(\mathbf{k}, N2) & \ldots & D(\mathbf{k}, NN) \end{pmatrix}, \tag{2.15}$$

and Eq. (2.13) is rewritten as

$$D(\mathbf{k})\epsilon_{\mathbf{k}\sigma} = \omega^2_{\mathbf{k}\sigma}\epsilon_{\mathbf{k}\sigma}. \tag{2.16}$$

The displacement polarization vector $\epsilon_{\mathbf{k}\sigma}$ satisfies the following orthogonality and completeness relations:

$$\epsilon^{\dagger}_{\mathbf{k}\sigma}\epsilon_{\mathbf{k}\sigma'} = \delta_{\sigma,\sigma'}, \tag{2.17}$$

$$\sum_{\sigma}\epsilon_{\mathbf{k}\sigma} \otimes \epsilon^{\dagger}_{\mathbf{k}\sigma} = I_{3N\times 3N}. \tag{2.18}$$

Here, we consider the time-reversal and inversion symmetries for lattice dynamics. We take the complex conjugate of Eq. (2.16) and use Eq. (2.11). Then we get

$$\epsilon^{\dagger}_{\mathbf{k}\sigma}D(\mathbf{k}) = \omega^2_{\mathbf{k}\sigma}\epsilon^{\dagger}_{\mathbf{k}\sigma} \tag{2.19}$$

From Eq. (2.16), for wavevector $-\mathbf{k}$, one has

$$D(-\mathbf{k})\epsilon_{-\mathbf{k}\sigma} = \omega^2_{-\mathbf{k}\sigma}\epsilon_{-\mathbf{k}\sigma}. \tag{2.20}$$

We then take the transpose as

$$\epsilon^{T}_{-\mathbf{k}\sigma}D(-\mathbf{k})^T = \omega^2_{-\mathbf{k}\sigma}\epsilon^{T}_{-\mathbf{k}\sigma}. \tag{2.21}$$

From Eqs. (2.11) and (2.12), the dynamical matrix $D$ satisfies $D^T(-\mathbf{k}) = D(\mathbf{k})$. Therefore, we get

$$\epsilon^{T}_{-\mathbf{k}\sigma}D(\mathbf{k}) = \omega^2_{-\mathbf{k}\sigma}\epsilon^{T}_{-\mathbf{k}\sigma}. \tag{2.22}$$

From Eqs. (2.19) and (2.22), we have

$$\omega^2_{\mathbf{k}\sigma} = \omega^2_{-\mathbf{k}\sigma}, \tag{2.23}$$

namely

$$\omega_{\mathbf{k}\sigma} = \omega_{-\mathbf{k}\sigma}, \tag{2.24}$$

$$\epsilon^{*}_{\mathbf{k}\sigma} = \epsilon_{-\mathbf{k}\sigma}. \tag{2.25}$$

This relation holds in systems with time-reversal symmetry. On the other hand, when the system has inversion symmetry, the displacement polarization vector has following relation:

$$\epsilon_{-\mathbf{k}\sigma} = \epsilon_{\mathbf{k}\sigma}. \tag{2.26}$$

### 2.1.2 Formalism of the Phonon Angular Momentum

In general crystals, the angular momenta of the atoms include both rigid-body rotations of crystals and microscopic local rotations of atoms around their equilibrium positions. The angular momentum associated with this microscopic local rotation is called phonon angular momentum [1], and it is represented as

$$\mathbf{J}^{\mathrm{ph}} = \sum_{l\kappa} \mathbf{u}_{l\kappa} \times m_{\kappa} \dot{\mathbf{u}}_{l\kappa}, \tag{2.27}$$

where $\mathbf{u}_{l\kappa}$ is a displacement vector of the $\kappa$th atom in the $l$th unit cell, and $m_{\kappa}$ is mass of $\kappa$th atom in unit cell. Let $N$ denote the number of atoms per unit cell. In terms of the displacement vectors $\mathbf{u}_{l\kappa}$, we define a vector $u_l$ associated with the $l$th unit cell with additional factors of the multiplied square root mass:

$$u_l = (\sqrt{m_1}u_{l1x}, \sqrt{m_1}u_{l1y}, \sqrt{m_1}u_{l1z}, \ldots, \sqrt{m_N}u_{lNz}, )^T. \tag{2.28}$$

The phonon angular momentum is rewritten as

$$J_{\alpha}^{\mathrm{ph}} = \sum_{l} u_l^T \mathrm{i} M_{\alpha} \dot{u}_l, \tag{2.29}$$

$$(M_{\alpha})_{\beta\gamma} = I_{N\times N} \otimes (-\mathrm{i})\varepsilon_{\alpha\beta\gamma}, \tag{2.30}$$

where the matrix $M_{\alpha}$ is a tensor product of the unit matrix for $N$ sites within the unit cell and a generator of the SO(3) rotation, represented by the anti-symmetric tensor $\varepsilon_{\alpha\beta\gamma}$. One can express the displacement vectors in the second quantization form as

$$u_l = \sum_{\mathbf{k}\sigma} \epsilon_{\mathbf{k}\sigma} e^{\mathrm{i}(\mathbf{R}_l\cdot\mathbf{k}-\omega_{\mathbf{k}\sigma}t)} \sqrt{\frac{\hbar}{2\omega_{\mathbf{k}\sigma}N}} a_{\mathbf{k}\sigma} + \epsilon_{\mathbf{k}\sigma}^{*} e^{-\mathrm{i}(\mathbf{R}_l\cdot\mathbf{k}-\omega_{\mathbf{k}\sigma}t)} \sqrt{\frac{\hbar}{2\omega_{\mathbf{k}\sigma}N}} a_{\mathbf{k}\sigma}^{\dagger}. \tag{2.31}$$

Substituting Eq. (2.31) into Eq. (2.29), we obtain

$$
\begin{aligned}
J_\alpha^{\rm ph} = \frac{\hbar}{2N}\sum_l\sum_{\mathbf{k},\mathbf{k}'}\sum_{\sigma,\sigma'}\Bigg[&\epsilon_{\mathbf{k}\sigma}^T M_\alpha \epsilon_{\mathbf{k}'\sigma'}\sqrt{\frac{\omega_{\mathbf{k}'\sigma'}}{\omega_{\mathbf{k}\sigma}}} a_{\mathbf{k}\sigma} a_{\mathbf{k}'\sigma'} e^{\mathrm{i}\mathbf{R}_l\cdot(\mathbf{k}+\mathbf{k}')-\mathrm{i}(\omega_{\mathbf{k}\sigma}+\omega_{\mathbf{k}'\sigma'})t} \\
&- \epsilon_{\mathbf{k}\sigma}^T M_\alpha \epsilon_{\mathbf{k}'\sigma'}^* \sqrt{\frac{\omega_{\mathbf{k}'\sigma'}}{\omega_{\mathbf{k}\sigma}}} a_{\mathbf{k}\sigma} a_{\mathbf{k}'\sigma'}^\dagger e^{\mathrm{i}\mathbf{R}_l\cdot(\mathbf{k}-\mathbf{k}')-\mathrm{i}(\omega_{\mathbf{k}\sigma}-\omega_{\mathbf{k}'\sigma'})t} \\
&+ \epsilon_{\mathbf{k}\sigma}^\dagger M_\alpha \epsilon_{\mathbf{k}'\sigma'}\sqrt{\frac{\omega_{\mathbf{k}'\sigma'}}{\omega_{\mathbf{k}\sigma}}} a_{\mathbf{k}\sigma}^\dagger a_{\mathbf{k}'\sigma'} e^{-\mathrm{i}\mathbf{R}_l\cdot(\mathbf{k}-\mathbf{k}')+\mathrm{i}(\omega_{\mathbf{k}\sigma}-\omega_{\mathbf{k}'\sigma'})t} \\
&- \epsilon_{\mathbf{k}\sigma}^\dagger M_\alpha \epsilon_{\mathbf{k}'\sigma'}^T \sqrt{\frac{\omega_{\mathbf{k}'\sigma'}}{\omega_{\mathbf{k}\sigma}}} a_{\mathbf{k}\sigma}^\dagger a_{\mathbf{k}'\sigma'}^\dagger e^{-\mathrm{i}\mathbf{R}_l\cdot(\mathbf{k}+\mathbf{k}')+\mathrm{i}(\omega_{\mathbf{k}\sigma}+\omega_{\mathbf{k}'\sigma'})t}\Bigg].
\end{aligned}
\tag{2.32}
$$

Here we ignore the $a_{\mathbf{k}\sigma} a_{\mathbf{k}'\sigma'}$ and $a_{\mathbf{k}\sigma}^\dagger a_{\mathbf{k}'\sigma'}^\dagger$ terms since they change rapidly with time and have no contribution in equilibrium. Thus, Eq. (2.32) is rewritten as

$$
\begin{aligned}
J_\alpha^{\rm ph} &= \frac{\hbar}{2N}\sum_l\sum_{\mathbf{k},\mathbf{k}'}\sum_{\sigma,\sigma'}\left(\epsilon_{\mathbf{k}\sigma}^\dagger M_\alpha \epsilon_{\mathbf{k}'\sigma'}\sqrt{\frac{\omega_{\mathbf{k}'\sigma'}}{\omega_{\mathbf{k}\sigma}}} a_{\mathbf{k}\sigma}^\dagger a_{\mathbf{k}'\sigma'}\right. \\
&\qquad\qquad \left. - \epsilon_{\mathbf{k}'\sigma'}^T M_\alpha \epsilon_{\mathbf{k}\sigma}^* \sqrt{\frac{\omega_{\mathbf{k}\sigma}}{\omega_{\mathbf{k}'\sigma'}}} a_{\mathbf{k}'\sigma'} a_{\mathbf{k}\sigma}^\dagger \right) e^{\mathrm{i}\mathbf{R}_l\cdot(\mathbf{k}-\mathbf{k}')-\mathrm{i}(\omega_{\mathbf{k}\sigma}-\omega_{\mathbf{k}'\sigma'})t} \\
&= \frac{\hbar}{2}\sum_{\mathbf{k}}\sum_{\sigma,\sigma'}\epsilon_{\mathbf{k}\sigma}^\dagger M_\alpha \epsilon_{\mathbf{k}\sigma'}\left(\sqrt{\frac{\omega_{\mathbf{k}\sigma'}}{\omega_{\mathbf{k}\sigma}}} a_{\mathbf{k}\sigma}^\dagger a_{\mathbf{k}\sigma'} + \sqrt{\frac{\omega_{\mathbf{k}\sigma}}{\omega_{\mathbf{k}\sigma'}}} a_{\mathbf{k}\sigma'} a_{\mathbf{k}\sigma}^\dagger\right) e^{-\mathrm{i}(\omega_{\mathbf{k}\sigma}-\omega_{\mathbf{k}\sigma'})t}
\end{aligned}
\tag{2.33}
$$

$$
\begin{aligned}
&= \frac{\hbar}{2}\sum_{\mathbf{k}}\sum_{\sigma,\sigma'}\epsilon_{\mathbf{k}\sigma}^\dagger M_\alpha \epsilon_{\mathbf{k}\sigma'}\left(\sqrt{\frac{\omega_{\mathbf{k}\sigma'}}{\omega_{\mathbf{k}\sigma}}} + \sqrt{\frac{\omega_{\mathbf{k}\sigma}}{\omega_{\mathbf{k}\sigma'}}}\right) a_{\mathbf{k}\sigma}^\dagger a_{\mathbf{k}\sigma'} e^{-\mathrm{i}(\omega_{\mathbf{k}\sigma}-\omega_{\mathbf{k}\sigma'})t} \\
&+ \frac{\hbar}{2}\sum_{\mathbf{k},\sigma}\epsilon_{\mathbf{k}\sigma}^\dagger M_\alpha \epsilon_{\mathbf{k}\sigma}.
\end{aligned}
\tag{2.34}
$$

In Eq. (2.33), we used $-\epsilon_{\mathbf{k}\sigma'}^T M_\alpha \epsilon_{\mathbf{k}\sigma}^* = \epsilon_{\mathbf{k}\sigma}^\dagger M_\alpha \epsilon_{\mathbf{k}\sigma'}$ and $\frac{1}{N}\sum_l e^{-\mathrm{i}\mathbf{R}_l\cdot(\mathbf{k}-\mathbf{k}')} = \delta_{\mathbf{k},\mathbf{k}'}$, and in Eq. (2.34), we used the commutation relation $[a_{\mathbf{k}\sigma'}, a_{\mathbf{k}\sigma}^\dagger] = \delta_{\sigma,\sigma'}$. In equilibrium we have $\langle a_{\mathbf{k}\sigma'}^\dagger a_{\mathbf{k}\sigma} =\rangle f_0(\omega_{\mathbf{k}\sigma})\delta_{\sigma,\sigma'}$, and we obtain

$$
J_\alpha^{\rm ph} = \hbar\sum_{\mathbf{k}\sigma}\epsilon_{\mathbf{k}\sigma}^\dagger M_\alpha \epsilon_{\mathbf{k}\sigma}\left(f_0(\omega_{\mathbf{k}\sigma}) + \frac{1}{2}\right), \quad l_{\mathbf{k}\sigma,\alpha} = \hbar\epsilon_{\mathbf{k}\sigma}^\dagger M_\alpha \epsilon_{\mathbf{k}\sigma}, \tag{2.35}
$$

where $f_0(\omega_{\mathbf{k}\sigma}) = (\exp(\hbar\omega_{\mathbf{k}\sigma}/k_B T) - 1)^{-1}$ is the Bose distribution function with the chemical potential $\mu = 0$. In Eq. (2.35), we sum over all the wavevectors in the first Brillouin zone and all the phonon modes. $l_{\mathbf{k}\sigma,\alpha}$ shows the phonon angular momentum of the mode $\sigma$ at the wavevector $\mathbf{k}$. At zero temperature, the total phonon angular momentum is $J_\alpha^{\rm ph}(T=0) = \sum_{\mathbf{k}\sigma}\frac{1}{2}l_{\mathbf{k}\sigma,\alpha}$, which means that each mode at wavevector $\mathbf{k}$ has a zero-point angular momentum in addition to a zero-point energy $\hbar\omega_{\mathbf{k}\sigma}/2$.

### 2.1.3 Time-Reversal and Inversion Symmetries

We explain properties of the phonon angular momentum in the presence of time-reversal or inversion symmetries. When the system has time-reversal symmetry, the phonon angular momentum has the following relation from Eq. (2.25)

$$l_{-\mathbf{k}\sigma,\alpha} = \hbar\epsilon_{-\mathbf{k}\sigma}^{\dagger} M_{\alpha}\epsilon_{-\mathbf{k}\sigma} = \hbar\epsilon_{\mathbf{k}\sigma}^{T} M_{\alpha}\epsilon_{\mathbf{k}\sigma}^{*} = \hbar\epsilon_{\mathbf{k}\sigma}^{\dagger} M_{\alpha}^{T}\epsilon_{\mathbf{k}\sigma} = -l_{\mathbf{k}\sigma,\alpha}. \tag{2.36}$$

Therefore, the phonon angular momentum for each mode is an odd function of the wavevector $\mathbf{k}$. From Eq. (2.24), we have $f_0(\omega_{-\mathbf{k}\sigma}) = f_0(\omega_{\mathbf{k}\sigma})$, and

$$\sum_{\mathbf{k}\sigma} l_{\mathbf{k}\sigma,\alpha}\left(f(\omega_{\mathbf{k}\sigma}) + \frac{1}{2}\right) = \sum_{\mathbf{k}'\sigma} l_{-\mathbf{k}'\sigma,\alpha}\left(f(\omega_{-\mathbf{k}'\sigma}) + \frac{1}{2}\right) = \sum_{\mathbf{k}\sigma} -l_{\mathbf{k}\sigma,\alpha}\left(f(\omega_{\mathbf{k}\sigma}) + \frac{1}{2}\right). \tag{2.37}$$

Therefore, the phonon angular momentum vanishes in systems with time-reversal symmetry.

On the other hand, when the system has inversion symmetry, the phonon angular momentum satisfies the following relation due to Eq. (2.26)

$$l_{-\mathbf{k}\sigma,\alpha} = \hbar\epsilon_{-\mathbf{k}\sigma}^{\dagger} M_{\alpha}\epsilon_{-\mathbf{k}\sigma} = \hbar\epsilon_{\mathbf{k}\sigma} M_{\alpha}\epsilon_{\mathbf{k}\sigma} = l_{\mathbf{k}\sigma,\alpha}. \tag{2.38}$$

Therefore, the phonon angular momentum for each mode is an even function of the wavevector $\mathbf{k}$. Thus, when the system has both time-reversal and inversion symmetries, the phonon angular momentum for each mode vanishes at all the wavevector $\mathbf{k}$.

### 2.1.4 Phonon System Without Time-Reversal Symmetry

We explain properties of phonons in systems without time-reversal symmetry [8–15]. Time-reversal symmetry can be broken by an external magnetic field. However, phonons in systems with neutral particles do not couple with a magnetic field directly because neutral particles do not have a charge. Indirect coupling between the phonons and the magnetic field is present in ionic crystals by the Lorentz force on charged ions, or in magnetic crystals by the Raman spin-phonon interaction [16–20]. Moreover, time-reversal symmetry can also be broken by the Coriolis force in rotating systems. As a related topic, in mechanical systems, the mechanical graphene in a rotating frame [21] and gyroscopic phononic crystals [22, 23] have been proposed theoretically as setups without time-reversal symmetry and have been realized experimentally. The Lagrangian for the phonons in the harmonic approximation is represented as

$$L_0 = \sum_l \frac{1}{2}\dot{u}_l^T \dot{u}_l - \frac{1}{2}\sum_{l,l'} u_l^T \Phi(l, l')u_{l'}, \tag{2.39}$$

with $u_l = (\sqrt{m_1}u_{l,1x}, \sqrt{m_1}u_{l,1y}, \sqrt{m_1}u_{l,1z}, \sqrt{m_2}u_{l,2x}, \ldots, \sqrt{m_N}u_{l,Nz})^T$. Next, we introduce a term which breaks the time-reversal symmetry [9]. In the harmonic approximation, the only physically allowed term that breaks the time-reversal symmetry is represented as

$$L' = \sum_{l\kappa} \eta_{ij}\dot{u}_{l\kappa,i}u_{l\kappa,j}, \tag{2.40}$$

where the matrix $\eta$ is real and antisymmetric. The Lagrangian for the phonons is rewritten as

$$L = \sum_l \frac{1}{2}\dot{u}_l^T \dot{u}_l - \frac{1}{2}\sum_{l,l'} u_l^T \Phi(l, l')u_{l'} + \sum_l \dot{u}_l^T \eta u_l \tag{2.41}$$

with $\eta = I_{Nd\times Nd} \otimes \eta_{ij}$. The canonical momentum is given by $p_l = \partial L/\partial \dot{u}_l = \dot{u}_l + \eta u_l$, and the Hamiltonian for the phonons $H = \sum_l p_l^T \dot{u}_l - L$ is written as

$$H = \sum_l \frac{1}{2}(p_l - \eta u_l)^T (p_l - \eta u_l) + \sum_{l,l'} u_l^T \Phi(l, l')u_l. \tag{2.42}$$

Then the equation of motion is obtained as

$$\ddot{u}_l + 2\eta\dot{u}_l + \sum_{l'} \Phi(l, l')u_{l'} = 0. \tag{2.43}$$

The displacement vector is represented by the displacement polarization vector in $k$-space,

$$u_l = \sum_{\mathbf{k}} \epsilon_{\mathbf{k}} e^{\mathrm{i}\mathbf{k}\cdot\mathbf{R}_l - \mathrm{i}\omega t}. \tag{2.44}$$

Then, the equation of the motion in $k$-space is written as

$$[(-\mathrm{i}\omega + \eta)^2 + D_{\mathbf{k}}]\epsilon_{\mathbf{k}} = 0, \tag{2.45}$$

$$D_{\mathbf{k}} = -\eta^2 + \sum_{l'} \Phi(l, l')e^{\mathrm{i}(\mathbf{R}_{l'} - \mathbf{R}_l)\cdot\mathbf{k}}, \tag{2.46}$$

where $D_{\mathbf{k}}$ is the Hermitian dynamical matrix in systems without time-reversal symmetry. Note that in phonon systems without time-reversal symmetry the phonon modes $\sigma$ include not only positive modes ($\sigma > 0$) but also negative modes ($\sigma < 0$),

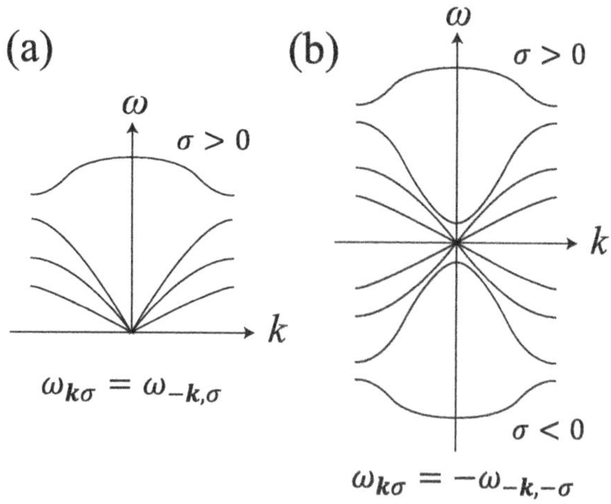


**Fig. 2.2** Schematic figure of the phonon dispersion in $k$-space **a** with the time-reversal symmetry and **b** without the time-reversal symmetry

and then the phonon dispersion is shown as Fig. 2.2. We take the complex conjugate of Eq. (2.45), and we get

$$\epsilon^{\dagger}_{\mathbf{k}\sigma}[(\mathrm{i}\omega_{\mathbf{k}\sigma} - \eta)^2 + D_{\mathbf{k}}] = 0. \tag{2.47}$$

From Eq. (2.45) for wavevector $-\mathbf{k}$ and phonon mode $-\sigma$, one has

$$[(-\mathrm{i}\omega_{-\mathbf{k},-\sigma} + \eta)^2 + D_{-\mathbf{k}}]\epsilon_{-\mathbf{k},-\sigma} = 0. \tag{2.48}$$

We then take the transpose as

$$\epsilon^{T}_{-\mathbf{k},-\sigma}[(-\mathrm{i}\omega_{-\mathbf{k},-\sigma} - \eta)^2 + D_{\mathbf{k}})] = 0. \tag{2.49}$$

From Eqs. (2.47) and (2.49), we have

$$\epsilon_{-\mathbf{k},-\sigma} = \epsilon^{*}_{\mathbf{k},\sigma}, \tag{2.50}$$

$$\omega_{-\mathbf{k},-\sigma} = -\omega_{\mathbf{k}\sigma}. \tag{2.51}$$

We note that Eq. (2.45) is not a standard eigenvalue problem. To obtain the phonon dispersion and the eigenvectors of the phonon modes, we need to solve Eq. (2.45). In this section, we rewrite this problem into a non-hermitian eigenvalue problem [11, 14]. We introduce an extended polarization vector $x = (\mu, \epsilon)^T$, where $\mu$ and $\epsilon$ are associated with the momenta and the displacement vector, respectively. The polarization vectors $\mu$ and $\epsilon$ satisfy the following relations:

$$-\mathrm{i}\omega\mu_{\mathbf{k}} = -\eta\mu_{\mathbf{k}} - D_{\mathbf{k}}\epsilon_{\mathbf{k}}, \tag{2.52}$$

$$-\mathrm{i}\omega\epsilon_{\mathbf{k}} = \mu_{\mathbf{k}} - \eta\epsilon_{\mathbf{k}}. \tag{2.53}$$

Then, we obtain

$$H_{\text{eff}} x_{\mathbf{k}\sigma} = \omega_{\mathbf{k}\sigma} x_{\mathbf{k}\sigma}, \tag{2.54}$$

$$\tilde{x}^T_{\mathbf{k}\sigma} H_{\text{eff}} = \omega_{\mathbf{k}\sigma} \tilde{x}_{\mathbf{k}\sigma}, \tag{2.55}$$

$$H_{\text{eff}} = \mathrm{i}\begin{pmatrix} -\eta & -D_{\mathbf{k}} \\ I & -\eta \end{pmatrix}, \tag{2.56}$$

where $x_{\mathbf{k}\sigma} = (\mu_{\mathbf{k}\sigma}, \epsilon_{\mathbf{k}\sigma})^T$ is the right eigenvector, and $\tilde{x}^T_{\mathbf{k}\sigma} = (\epsilon^\dagger_{\mathbf{k}\sigma}, -\mu^\dagger_{\mathbf{k}\sigma})/(-2\mathrm{i}\omega_{\mathbf{k}\sigma})$ is the left eigenvector. Since the effective Hamiltonian $H_{\text{eff}}$ is not Hermitian, the orthonormal condition holds between the left and the right eigenvectors as

$$\tilde{x}^T_{\mathbf{k}\sigma} x_{\mathbf{k}\sigma'} = \delta_{\sigma,\sigma'}. \tag{2.57}$$

We also have the completeness relation

$$\sum_\sigma x_{\mathbf{k}\sigma} \otimes \tilde{x}^T_{\mathbf{k}\sigma} = I_{2Nd}. \tag{2.58}$$

The normalization of the eigenmodes is equivalent to

$$\epsilon^\dagger_{\mathbf{k}\sigma}\epsilon_{\mathbf{k}\sigma} + \frac{\mathrm{i}}{\omega_{\mathbf{k}\sigma}}\epsilon^\dagger_{\mathbf{k}\sigma}\eta\epsilon_{\mathbf{k}\sigma} = 1. \tag{2.59}$$

Next, we show the phonon angular momentum without time-reversal symmetries. We substitute the equation of motion $\dot{u}_l = p_l - \eta u_l$ into the phonon angular momentum Eq. (2.29), and the phonon angular momentum is rewritten as

$$\begin{aligned} J^{\text{ph}}_\alpha &= \sum_l u_l^T \mathrm{i} M_\alpha p_l - u_l^T \mathrm{i} M_\alpha \eta u_l \\ &= \sum_l \begin{pmatrix} u_l \\ -p_l \end{pmatrix}^T \begin{pmatrix} \mathrm{i}M_\alpha & -\mathrm{i}M_\alpha\eta \\ O & O \end{pmatrix} \begin{pmatrix} p_l \\ u_l \end{pmatrix} \\ &= \sum_l \tilde{\chi}_l^T \begin{pmatrix} \mathrm{i}M_\alpha & -\mathrm{i}M_\alpha\eta \\ O & O \end{pmatrix} \chi_l, \end{aligned} \tag{2.60}$$

with $\chi_l = (p_l, u_l)^T$ and $\tilde{\chi}_l^T = (u_l, -p_l)$. By second quantization

$$\chi_l = \sqrt{\frac{\hbar}{N}} \sum_{\mathbf{k},\sigma} x_{\mathbf{k}\sigma} e^{\mathrm{i}\mathbf{R}_l\cdot\mathbf{k}} \sqrt{\frac{1}{2|\omega_{\mathbf{k}\sigma}|}} a_{\mathbf{k}\sigma}, \tag{2.61}$$

$$\tilde{\chi}_l^T = \sqrt{\frac{\hbar}{N}} \sum_{\mathbf{k},\sigma} \tilde{x}^T_{\mathbf{k}\sigma} e^{-\mathrm{i}\mathbf{R}_l\cdot\mathbf{k}} (-2\omega_{\mathbf{k}\sigma}) \sqrt{\frac{1}{2|\omega_{\mathbf{k}\sigma}|}} a^\dagger_{\mathbf{k}\sigma}, \tag{2.62}$$

we obtain

$$J_\alpha^{\mathrm{ph}} = \frac{\hbar}{N}\sum_{l,\mathbf{k},\sigma,\mathbf{k}',\sigma'} \left(\epsilon_{\mathbf{k}\sigma}^\dagger \; -\mu_{\mathbf{k}\sigma}^\dagger\right)\begin{pmatrix} \mathrm{i}M_\alpha & -\mathrm{i}M_\alpha\eta \\ O & O\end{pmatrix}\begin{pmatrix}\mu_{\mathbf{k}'\sigma'} \\ \epsilon_{\mathbf{k}'\sigma'}\end{pmatrix}\frac{1}{2\sqrt{|\omega_{\mathbf{k}\sigma}||\omega_{\mathbf{k}'\sigma'}|}}a_{\mathbf{k}\sigma}^\dagger a_{\mathbf{k}'\sigma'}$$

$$= \frac{\hbar}{2}\sum_{\mathbf{k},\sigma,\sigma'} \left(\epsilon_{\mathbf{k}\sigma}^\dagger \; -\mu_{\mathbf{k}\sigma}^\dagger\right)\begin{pmatrix} \mathrm{i}M_\alpha & -\mathrm{i}M_\alpha\eta \\ O & O\end{pmatrix}\begin{pmatrix}\mu_{\mathbf{k}\sigma'} \\ \epsilon_{\mathbf{k}\sigma'}\end{pmatrix}\frac{1}{\sqrt{|\omega_{\mathbf{k}\sigma}||\omega_{\mathbf{k}\sigma'}|}}a_{\mathbf{k}\sigma}^\dagger a_{\mathbf{k}\sigma'}. \quad (2.63)$$

In equilibrium, because $\langle a_{\mathbf{k}\sigma}^\dagger a_{\mathbf{k}\sigma'} =\rangle f_0(\omega_{\mathbf{k}\sigma})\mathrm{sign}(\sigma)\delta_{\sigma,\sigma'}$, its expectation value is given by

$$J_\alpha^{\mathrm{ph}} = \frac{\hbar}{2}\sum_{\mathbf{k}\sigma} \left(\epsilon_{\mathbf{k}\sigma}^\dagger \; -\mu_{\mathbf{k}\sigma}^\dagger\right)\begin{pmatrix} \mathrm{i}M_\alpha & -\mathrm{i}M_\alpha\eta \\ O & O\end{pmatrix}\begin{pmatrix}\mu_{\mathbf{k}\sigma'} \\ \epsilon_{\mathbf{k}\sigma'}\end{pmatrix}\frac{f_0(\omega_{\mathbf{k}\sigma})\mathrm{sign}(\sigma)}{|\omega_{\mathbf{k}\sigma}|}. \quad (2.64)$$

Here using the equation of motion in k-space $\mu_{\mathbf{k}\sigma} = -\mathrm{i}\omega_{\mathbf{k}\sigma}\epsilon_{\mathbf{k}\sigma} + \eta\epsilon_{\mathbf{k}\sigma}$, we obtain

$$\left(\epsilon_{\mathbf{k}\sigma}^\dagger \; -\mu_{\mathbf{k}\sigma}^\dagger\right)\begin{pmatrix} \mathrm{i}M_\alpha & -\mathrm{i}M_\alpha\eta \\ O & O\end{pmatrix}\begin{pmatrix}\mu_{\mathbf{k}\sigma'} \\ \epsilon_{\mathbf{k}\sigma'}\end{pmatrix} = \omega_{\mathbf{k}\sigma}\epsilon_{\mathbf{k}\sigma}^\dagger M_\alpha\epsilon_{\mathbf{k}\sigma}. \quad (2.65)$$

The phonon angular momentum in equilibrium is represented as

$$J_\alpha^{\mathrm{ph}} = \frac{\hbar}{2}\sum_{\mathbf{k}\sigma}\epsilon_{\mathbf{k}\sigma}^\dagger M_\alpha\epsilon_{\mathbf{k}\sigma} f_0(\omega_{\mathbf{k}\sigma}) \quad (2.66)$$

by using $\omega_{\mathbf{k}\sigma}\mathrm{sign}(\sigma) = |\omega_{\mathbf{k}\sigma}|$. Thus the completed set of this non-hermitian eigenvalues problem contains the phonon modes of positive and negative frequencies. From Eq. (2.50) , we get

$$\epsilon_{-\mathbf{k},-\sigma}^\dagger M_\alpha\epsilon_{-\mathbf{k},-\sigma} = \epsilon_{\mathbf{k}\sigma}^T M_\alpha\epsilon_{\mathbf{k}\sigma}^* = -\epsilon_{\mathbf{k}\sigma}^\dagger M_\alpha\epsilon_{\mathbf{k}\sigma}. \quad (2.67)$$

Meanwhile, because $f_0(\omega_{-\mathbf{k},-\sigma}) \neq f_0(\omega_{\mathbf{k}\sigma})$, the phonon angular momentum at $(-\mathbf{k}, -\sigma)$ does not cancel that at $(\mathbf{k}, \sigma)$, and we can get a nonzero phonon angular momentum. For the negative phonon modes, the phonon angular momentum is rewritten as

$$\frac{\hbar}{2}\sum_{\mathbf{k},\sigma<0}\epsilon_{\mathbf{k}\sigma}^\dagger M_\alpha\epsilon_{\mathbf{k}\sigma} f_0(\omega_{\mathbf{k}\sigma}) = \frac{\hbar}{2}\sum_{\mathbf{k}',\sigma'>0}\epsilon_{-\mathbf{k}',-\sigma'}^\dagger M_\alpha\epsilon_{-\mathbf{k}',-\sigma'} f_0(\omega_{-\mathbf{k}',-\sigma'})$$

$$= \frac{\hbar}{2}\sum_{\mathbf{k}',\sigma'>0}\epsilon_{\mathbf{k}'\sigma'}^T M_\alpha\epsilon_{\mathbf{k}'\sigma'}^* f_0(-\omega_{\mathbf{k}'\sigma'})$$

$$= \frac{\hbar}{2}\sum_{\mathbf{k}',\sigma'>0}\epsilon_{\mathbf{k}'\sigma'}^\dagger(-M_\alpha)\epsilon_{\mathbf{k}'\sigma'}(-1 - f(\omega_{\mathbf{k}'\sigma'}))$$

$$= \frac{\hbar}{2} \sum_{\mathbf{k}',\sigma'>0} \epsilon^{\dagger}_{\mathbf{k}'\sigma'} M_\alpha \epsilon_{\mathbf{k}'\sigma'} (1 + f(\omega_{\mathbf{k}'\sigma'})). \tag{2.68}$$

In the third equality, we use $\epsilon^T_{\mathbf{k}',\sigma'} M_\alpha \epsilon^*_{\mathbf{k}'\sigma'} = \epsilon^{\dagger}_{\mathbf{k}'\sigma'} M^T_\alpha \epsilon_{\mathbf{k}'\sigma'}$, $M^T_\alpha = -M_\alpha$, and $f_0(-\omega_{\mathbf{k}'\sigma'}) = -(1 + f_0(\omega_{\mathbf{k}'\sigma'}))$. Therefore, we get the phonon angular momentum in a system without time-reversal symmetry as

$$J^{\mathrm{ph}}_\alpha = \sum_{\mathbf{k},\sigma>0} l_{\mathbf{k}\sigma,\alpha} \left( f_0(\omega_{\mathbf{k}\sigma}) + \frac{1}{2} \right). \tag{2.69}$$

### 2.1.5 Phonon Angular Momentum in the High- and Low-Temperature Limits

We calculate the phonon angular momentum in systems without time-reversal symmetry in the high- and the low-temperature limits [1]. First, in the high-temperature limit ($\hbar\omega_{\mathbf{k}\sigma} \ll k_{\mathrm{B}}T$), the Bose distribution function $f_0$ can be expanded in Taylor series as

$$f_0(x) = \frac{1}{e^x - 1} \simeq \frac{1}{x} - \frac{1}{2} + \frac{1}{12}x + O(x^2), \quad x = \frac{\hbar\omega}{k_{\mathrm{B}}T}. \tag{2.70}$$

Substituting it into Eq. (2.69), we get

$$J^{\mathrm{ph}}_\alpha = \sum_{\mathbf{k},\sigma>0} l_{\mathbf{k}\sigma,\alpha} \left( \frac{k_{\mathrm{B}}T}{\hbar\omega_{\mathbf{k}\sigma}} + \frac{\hbar\omega_{\mathbf{k}\sigma}}{12k_{\mathrm{B}}T} \right). \tag{2.71}$$

The first term is rewritten as

$$\begin{aligned} \sum_{\mathbf{k},\sigma>0} \frac{l_{\mathbf{k}\sigma,\alpha}}{\omega_{\mathbf{k}\sigma}} &= \sum_{\mathbf{k},\sigma>0} \frac{\epsilon^{\dagger}_{\mathbf{k}\sigma} M_\alpha \epsilon_{\mathbf{k}\sigma}}{\omega_{\mathbf{k}\sigma}} = \frac{1}{2} \sum_{\mathbf{k},\sigma} \sum_{i,j} \frac{\epsilon^*_{\mathbf{k}\sigma,i} M_{\alpha,ij} \epsilon_{\mathbf{k}\sigma,j}}{\omega_{\mathbf{k}\sigma}} \\ &= \frac{1}{2} \sum_{\mathbf{k},ij} M_{\alpha,ij} \sum_{\sigma} \frac{\epsilon^*_{\mathbf{k}\sigma,i} \epsilon_{\mathbf{k}\sigma,j}}{\omega_{\mathbf{k}\sigma}} = 0. \end{aligned} \tag{2.72}$$

In the second equality, we used the following relation

$$\sum_{\mathbf{k},\sigma<0} \frac{l_{\mathbf{k}\sigma,\alpha}}{\omega_{\mathbf{k}\sigma}} = \sum_{\mathbf{k}',\sigma'>0} \frac{l_{-\mathbf{k}',-\sigma'}}{\omega_{-\mathbf{k}',-\sigma'}} = \sum_{\mathbf{k}',\sigma'>0} \frac{-l_{\mathbf{k}',\sigma'}}{-\omega_{\mathbf{k}'\sigma'}} = \sum_{\mathbf{k},\sigma>0} \frac{l_{\mathbf{k}\sigma,\alpha}}{\omega_{\mathbf{k}\sigma}}, \tag{2.73}$$

and in the forth equality, we used the following relation from the completeness relation,

$$\sum_{\sigma} \frac{\epsilon_{\mathbf{k}\sigma} \otimes \tilde{\epsilon}^{\dagger}_{\mathbf{k}\sigma}}{-2i\omega_{\mathbf{k}\sigma}} = O_{3N \times 3N}. \tag{2.74}$$

Therefore, in the high-temperature limit, the phonon angular momentum is rewritten as

$$J^{\mathrm{ph}}_{\alpha} = \sum_{\mathbf{k},\sigma>0} \frac{\hbar\omega_{\mathbf{k}\sigma}}{12k_{\mathrm{B}}T} l_{\mathbf{k}\sigma,\alpha}, \tag{2.75}$$

and this converges to zero when $T \to \infty$. On the other hand, in the low-temperature limit, the Bose distribution function $f_0$ converges to zero, and thus the zero-point phonon angular momentum is dominant.

### 2.1.6 Chiral Phonons in Monolayer Hexagonal Lattices

We introduce the chirality of valley phonons [2]. When the phonon dispersion has local maxima or minima in the momentum space, they are called valleys. The phonons at the valley wave vectors are called as valley phonons. We focus on the two-dimensional honeycomb-lattice model, whose unit cell has two atoms A and B in an unit cell. The chiral phonon is related to the phonon angular momentum in systems without inversion symmetry. The honeycomb lattice, which has non-equivalent atoms A and B in an unit cell, is useful as a simplified model to demonstrate general features of chiral phonons. Therefore, chiral phonons can be realized e.g. in a graphene with a phononic gap due to isotopic doping [24], or staggered sublattice potential which is similar to hexagonal boron nitride [25].

To demonstrate the relation between chiral phonons and phonon angular momentum, we performed a numerical calculation based on a spring-mass model forming a honeycomb lattice as shown in Fig. 2.3a. We take the longitudinal spring constant $K_{\mathrm{L}} = 1$, the transverse one $K_{\mathrm{T}} = 0.25$, and the mass of atom A $m_{\mathrm{A}} = 1$, and the mass of atom B $m_{\mathrm{B}} = 1.2$. The primitive vectors are $(a, 0)$ and $(a/2, \sqrt{3}a/2)$ with the lattice constant $a$ and we show the first Brillouin zone in Fig. 2.3b. We show the phonon dispersion on the high-symmetry lines in Fig. 2.3c, and the phonon angular momentum of each atom, and the phonon angular momentum for each mode in Fig. 2.3d–f. In Fig. 2.3g, we show the trajectory of the atoms A and B for each phonon mode at $K$ point. In this model the phonon dispersion have local maxima or minima at $K$ and $K'$ points, and we associate the $K$ and $K'$ points with a valley degree of freedom. At the valleys located at $K$ and $K'$ points, all of the phonon modes are circularly polarized. From Eq. (2.36), the direction of circular polarization is opposite between the two valleys at $K$ and $K'$ points.

The phonon chirality can be characterized by the polarization of phonons. This mechanism is similar to the circularly polarized light. To consider the polarization along the $z$-direction, we introduce the eigenvector for phonons within the $xy$-plane

$\epsilon = (x_A, y_A, x_B, y_B)$. We introduce a new basis representing the right-handed and the left-handed circular polarization for each sublattice as $|R_A\rangle = \frac{1}{\sqrt{2}}(1, i, 0, 0)^T$, $|L_A\rangle = \frac{1}{\sqrt{2}}(1, -i, 0, 0)^T$, $|R_B\rangle = \frac{1}{\sqrt{2}}(0, 0, 1, i)^T$, and $|L_B\rangle = \frac{1}{\sqrt{2}}(0, 0, 1, -i)^T$, and the phonon eigenvector $\epsilon$ can be rewritten as

$$\epsilon = \sum_{n=A,B} \epsilon_{R_n} |R_n\rangle + \epsilon_{L_n} |L_n\rangle , \tag{2.76}$$

where $\epsilon_{R_n} = \langle R_n|\epsilon =\rangle \frac{1}{\sqrt{2}}(x_n - iy_n)$, and $\epsilon_{L_n} = \langle L_n|\epsilon =\rangle \frac{1}{\sqrt{2}}(x_n + iy_n)$. The operator for phonon circular polarization along the $z$-direction can be defined as

$$\hat{S}_z \equiv \sum_{n=A,B} (|R_n\rangle \langle R_n| - |L_n\rangle \langle L_n|), \tag{2.77}$$

and the phonon circular polarization is equal to

$$s_z^{\mathrm{ph}} = \epsilon^\dagger \hat{S}_z \epsilon \hbar = \sum_{n=A,B} (|\epsilon_{R_n}|^2 - |\epsilon_{L_n}|^2)\hbar. \tag{2.78}$$

The phonon circular polarization takes a value between $\pm\hbar$ since $\sum_n(|\epsilon_{R_n}|^2 + |\epsilon_{L_n}|^2) = 1$. The $s_z^{\mathrm{ph}}$ has the same form with that of phonon angular momentum for each mode along the $z$-direction. In Fig. 2.3d and e, at the valley $K$, $s_z^A = 0$, $s_z^B = -\hbar$ for the second lowest band, while $s_z^A = \hbar$, $s_z^B = 0$ for the third lowest band, where the phonon circular polarization happens to be quantized. For the first and forth lowest bands, the atoms A and B have opposite circular vibrations with different magnitudes, and the phonon circular polarizations have a nonzero value at the valley $K$ and $K'$ points. On the other hand, at the $\Gamma$ point, there are doubly degenerate acoustic modes and doubly degenerate optical modes, which are not circularly polarized in Fig. 2.3d, e. Here, we define the phonon pseudoangular momentum (PAM). In a honeycomb lattice, at high-symmetry points $\Gamma$, $K$, $K'$, phonons are invariant under a three-fold rotation around the $z$-direction (perpendicular to the plane). Under the three-fold rotation, one can obtain $R(2\pi/3, z)\mathbf{u_k} = e^{-i(2\pi/3)l_{\mathbf{k}}^{\mathrm{ph}}}\mathbf{u_k} = e^{-i(2\pi/3)(l_{\mathbf{k}}^s + l_{\mathbf{k}}^o)}\mathbf{u_k}$, where $l_{\mathbf{k}}^{\mathrm{ph}}$ is defined as the PAM of a phonon with a wave function $\mathbf{u_k}$ and has values of $\pm 1$ or 0. The change of the phase under the rotation of the phonon wave function consists of two parts: one is from the local (intracell) part $\epsilon_{\mathbf{k}\sigma}$ called spin PAM $l^s$, another is from the nonlocal (intercell) part $e^{i\mathbf{R}_l \cdot \mathbf{k}}$, called orbital PAM $l^o$. The orbital PAM is obtained from a phase change under a three-fold rotation, and at valley $K(K')$ point, we get $l_A^o = +1(-1)$, $l_B^o = -1(+1)$. At $\Gamma$ point, we obtain $l_A^o = l_B^o = 0$ since there is no phase change. On the other hand, the spin PAM at the valley $K(K')$ point is $l_A^s = -1(+1)$, $l_B^s = 1(-1)$ for the first and forth lowest bands, $l_B^s = -1(+1)$ for the second lowest band, and $l_A^s = +1(-1)$ for the third lowest band, since both circular polarizations are eigenstates of the operator $R(2\pi/3, z)$, with $l_R^s = 1$, $l_L^s = -1$, respectively. The PAM of the phonon equals $l^{\mathrm{ph}} = l_A^s + l_A^o = l_B^s + l_B^o$. Therefore, at

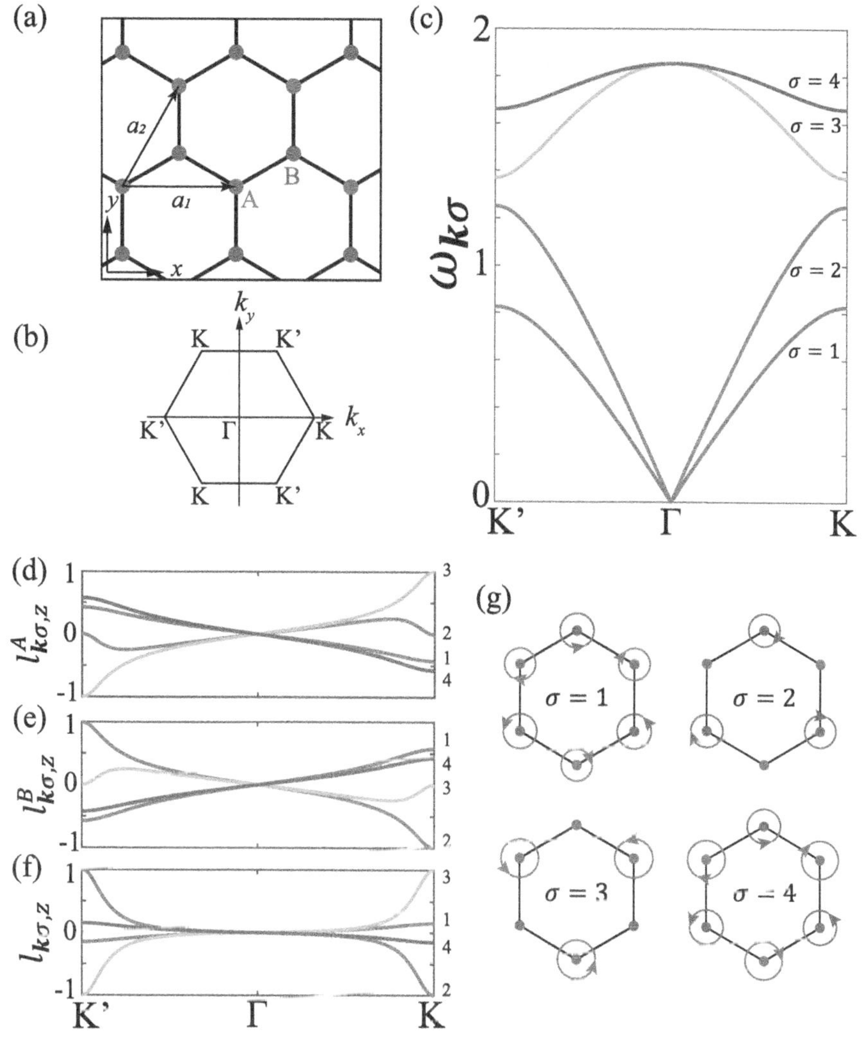


**Fig. 2.3** Phonon dispersion and the phonon angular momentum in the honeycomb lattice without inversion symmetry. **a** Lattice structure. **b** First Brillouin zone. **c** Phonon dispersion in the honeycomb lattice without inversion symmetry. Phonon angular momentum of **d** the sublattice A, of **e** the sublattice B, and **f** their sum. In **c**–**f** the colors represent the eigenmodes. **g** Trajectory of the sublattices A and B for each phonon mode at K point. Here, we take the parameters as the longitudinal spring constant $K_L = 1$, the transverse one $K_T = 0.25$, and $m_A = 1$, $m_B = 1.2$. The primitive vectors are $(a, 0)$ and $(a/2, \sqrt{3}a/2)$

the valley $K(K')$ point, $l^{\text{ph}}_{K(K')} = 1(-1)$ for the second lowest band, $l^{\text{ph}}_{K(K')} = -1(+1)$ for the third lowest band, and $l^{\text{ph}}_{K,K'} = 0$ for the first and fourth lowest bands.

Next, we explain the theoretical proposals and the experimental reports to observe chiral phonon. In the intervalley scattering involving phonons in a system with three-fold rotational symmetry, we can expect a selection rule from the conservation of the PAM. When the electrons or holes are scattered from $K$ point to $K'$ point, a circularly polarized phonon is emitted due to the conservation of the PAM: $l_K^{c(v)} - l_{K'}^{c(v)} = \pm 1$, while preserving momentum and energy. Thus, a valley phonon with PAM can be created by electrons or holes. For example, we assume that the phonons at valley $K$ of the lowest band is excited by electrons. Since the PAM of the valley phonon of each mode is different, through a left-circularly-polarized infrared photon with energy $\hbar\omega_{K,2} - \hbar\omega_{K,1}$ or a right-circularly-polarized infrared photon with energy $\hbar\omega_{K,3} - \hbar\omega_{K,1}$, where $\omega_{\mathbf{k}\sigma}$ is the eigenfrequency of each phonon mode as shown in Fig. 2.3c, one can observe a left- or a right-handed photoluminesence accordingly. Thus, from the resonance peak in the circularly polarized infrared spectrum, we can distinguish the valley phonons. Moreover, the observation of a chiral phonon has been reported for monolayer tungsten diselenide ($WSe_2$) in experiments [3]. When a left-handed circularly polarized photon is injected to $WSe_2$, a right-handed circularly polarized photon is emitted by the Raman process. Because holes in one valley generated by left-handed circularly polarized photons are scattered to the opposite valley and chiral phonons are generated, the polarization of emitting photon is reversed. Therefore, it was experimentally shown that chiral phonons are generated by circularly polarized photons.

## 2.2 Cross-Correlated Response

In this section, we give an overview of the relation between the macroscopic physical properties and crystallographic symmetry. In particular, we focus on the physical phenomena of cross correlation between different types of physical degrees of freedom, such as the induction of a magnetization (polarization) by an electric (magnetic) field as shown in Fig 2.4. Here, we discuss physical properties of a crystal given by responses of measurable quantities to an external field. Moreover, we introduce examples of the cross-correlated responses such as the Edelstein effect and the magnetoelectric effect. First, we explain that the point-group symmetry of the system determines which components of the response tensor can be nonzero [26–28].

Here, we introduce a concept of an axial and a polar tensor. We first write the coordinate transformation $\mathbf{x} \to \mathbf{x}'$, $x'_i = a_{ij}x_j$ with $a_{ij}$ being an orthogonal matrix. Then, a vector $\mathbf{v}$ is transformed as $\mathbf{v} \to \mathbf{v}'$, $v'_i = a_{ij}v_j$. Similarly, a tensor of rank two or higher is written as $T'_{ijk\ldots} = a_{im}a_{jn}a_{kl}\ldots T_{mnl\ldots}$. These relations mean that these quantities are invariant under the coordinate transformation. Now let us consider a vector $\mathbf{C} = \mathbf{A} \times \mathbf{B}$, where $\mathbf{A}$ and $\mathbf{B}$ are vectors. When we perform rotational operations to $\mathbf{C}$, $\mathbf{C}$ transforms like a vector. However, we perform an operation which converts from the right-handed to left-handed coordinate, such as mirror operation, we need to multiply $-1$ to the vector $\mathbf{C}$ to the transformation rule. This relation is written as $C'_i = \det(a)a_{ij}C_j$. Such vectors and tensors which require the additional

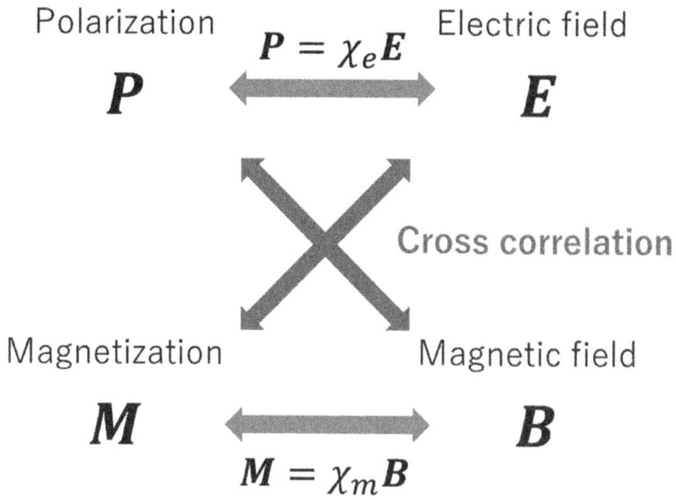


**Fig. 2.4** Schematic figure of the cross correlation response

det$(a)$ factor are called axial vectors and tensors. On the other hand, the vectors and tensors, which do not need this det$(a)$ factor, are called polar vectors and tensors. The form of a tensor representing physical properties of a crystal is restricted by the crystal symmetry. This relation is called Neumann's principle, which states that any type of symmetry described by the point group of the crystal is possessed by physical properties of the crystal. Let us introduce a response tensor $C$ representing the relation between a response $B$ and an external field $A$: $B_i = C_{ij} A_j$ in the crystal with point group $G$. If the operation $R$ represented by an orthogonal matrix $a_{ij}$ is the element of the point group $G$, the response tensor is transformed as

$$C'_{ij} = a_{ik} a_{jl} C_{kl} = C_{ij}, \tag{2.79}$$

if it is a polar tensor, and

$$C'_{ij} = \det(a) a_{ik} a_{jl} C_{kl} = C_{ij}, \tag{2.80}$$

if it is an axial tensor. By using this relationship, the form of the response tensor for the point group considered has been determined.

In many cases, we use crystallographic symmetry to determine macroscopic physical properties, and we do not need to consider the time-reversal symmetry. However, it is not sufficient for magnetic crystals with space-time symmetry. To describe such cases we introduce magnetic point groups. There are three types of magnetic point groups. Type 1 is ordinary point groups which do not include the time-reversal symmetry, and this type is used for describing a ferromagnet. A magnetic point group of type 2 consists of a set of an ordinary point group $G$ and a set of the same point group multiplied by the time-reversal operator TR TR $\times$ $G$, and it is used for describing paramagnets and diamagnets. In a magnetic point group of type 3, the time-reversal operator itself is not an element of the group. Instead a magnetic point group $M$ of type 3 is given by

$$M = H + \mathrm{TR} \times (G - H), \tag{2.81}$$

where $H$ is a halving subgroup of an ordinary point group $G$, including a half of the elements of $G$. Here, we prove that $M$ is a group. The point group $G$ has $h$ elements $\{A_i\}$ $(i = 1, 2, 3, \ldots, h)$, and the halving subgroup $H$ has $m$ elements $A_j$ $(j = 1, 2, 3, \ldots, m)$ and the rest has $(h - m)$ elements $\mathrm{TR} \times \{A_k\}$ $(k = m + 1, m + 2, \ldots, h)$. The point group is written as $G = \{A_j\} + \{A_k\}$, and $M$ is written as $M = \{A_j\} + \mathrm{TR} \times \{A_k\}$. Since $H$ is the subgroup of point group $G$, the product of $A_j(\mathrm{TR} \times A_k)$ is the elements of $\mathrm{TR} \times \{A_k\}$, and the product of $(\mathrm{TR} \times A_k)(\mathrm{TR} \times A_{k'})$ is the elements of $H$. Moreover, the product of one element of $\mathrm{TR} \times \{A_k\}$ and all the elements of $\{A_j\}$ creates the different $m$ elements of $\mathrm{TR} \times \{A_k\}$, and the product of one element $\mathrm{TR} \times A_k$ and all the elements of $\mathrm{TR} \times \{A_k\}$ creates the different $(h - m)$ elements of $\{A_j\}$. Then, if follows that $h - m = m$, and $H$ is a halving subgroup of the point group $G$. Thereby, $M$ itself is also a group. There are 58 magnetic point groups of type 3. Hence, there are 122 magnetic point groups.

Next, we explain the relationship between the macroscopic response tensor and the magnetic point group. Note that time-reversal-symmetric tensors and time-reversal-antisymmetric tensors must be distinguished in using the Neumann's principle. For type 1, describing a ferromagnet, since the time-reversal operator TR is not a symmetry operator, the forms of both time-reversal-symmetric and time-reversal-antisymmetric tensors take the same forms as those for an ordinary point group. For type 2, describing a paramagnet and a diamagnetic crystal, since the time-reversal operator TR is a symmetry operator, all the time-reversal-antisymmetric tensors vanish identically and all the time-reversal-symmetric tensors take the same form as those for an ordinary point group. For type 3, describing a multiferroic material, it is necessary to distinguish not only between polar and axial tensors but also between time-reversal-symmetric and time-reversal-antisymmetric tensors. The time-reversal-symmetric tensors take the same form as those for an ordinary point group. Meanwhile, the time-reversal-antisymmetric tensors obey following relations

$$C'_{ij} = (-1)a_{ik}a_{jl}C_{kl} = C_{ij}, \tag{2.82}$$

if it is a polar tensor, and

$$C'_{ij} = (-1)\det(a)a_{ik}a_{jl}C_{kl} = C_{ij}, \tag{2.83}$$

it it is an axial tensor from the Neumann's principle. Thus, we have shown how the response tensors for all magnetic point groups are determined.

Next, we introduce the Edelstein effect and the magnetoelectric effect as examples to show how the form of the response tensors are determined from symmetry.

### 2.2.1 Edelstein Effect

Spin polarization can be generated by applying an electric field to a metal with spin-split bands by the spin-orbit interaction. This effect is called the Edelstein effect [29–32]. We show the conceptual diagram of the Edelstein effect in Fig. 2.5. On surfaces or interfaces with the Rashba effect such as the Bi/Ag surface, due to the breaking of inversion symmetry, the band structure has a spin splitting coming from the spin-orbit interaction as shown in Fig. 2.5a. In Fig. 2.5b, we show the electron spin texture on the Fermi surface. In systems with the time-reversal symmetry but without inversion symmetry, the electron spins obey the relation $\mathbf{s}(k_x, k_y) = -\mathbf{s}(-k_x, -k_y)$, and therefore, in calculating the total spins, cancellation occurs between $(k_x, k_y)$ and $(-k_x, -k_y)$. When the electric field along the $-x$-direction is applied, the Fermi surface is slightly shifted to the $x$-direction, and the charge current with spin-polarized in the $y$-direction flows in the $x$-direction. The spin-polarization is written as

$$S_i = \alpha_{ij} E_j, \tag{2.84}$$

where $\mathbf{S}$ is the spin expectation value, $\mathbf{E}$ is the electric field, and $\alpha$ is response tensor. This tensor is an axial tensor, and its form is determined by the point group. In this example, this system has four-fold rotational symmetry along the $z$-direction and the mirror reflection symmetry with respect to the plane including the $z$ axis. Therefore, the response tensor $\alpha$ is written as

$$\alpha = \begin{pmatrix} 0 & \alpha_{xy} & 0 \\ -\alpha_{xy} & 0 & 0 \\ 0 & 0 & 0 \end{pmatrix}, \tag{2.85}$$

and this tensor is consistent with the result of Fig. 2.5c.

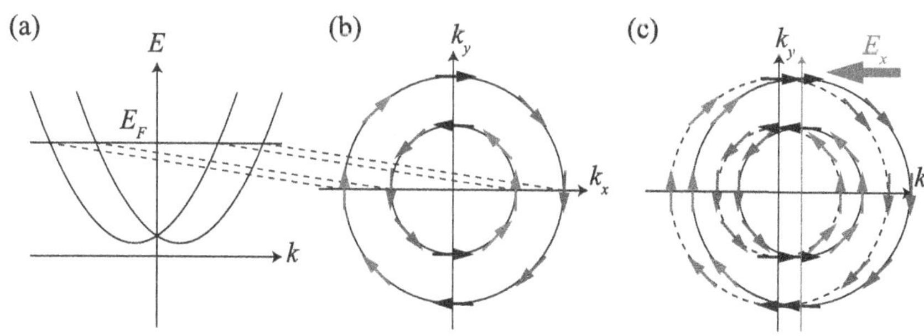


**Fig. 2.5** Conceptual diagram of the Edelstein effect. **a** Band structure for spin-split bands due to the Rashba spin-orbit interaction. **b** Electron spin texture on the Fermi surfaces. **c** Edelstein effect. The dashed circles show the Fermi surface in equilibrium and the solid circles show the Fermi surface with the electric field $-E_x$

### 2.2.2 *Magnetoelectric Effect*

Another example of physical phenomena of cross-correlation between different types of physical degrees of freedom is the magnetoelectric effect. In the magnetoelectric effect, an electric field induces a magnetization and a magnetic field induces a polarization. This effect occurs only in insulators. The magnetoelectric effect is defined as

$$\mathbf{M} = \alpha^T \mathbf{E}, \tag{2.86}$$

$$\mathbf{P} = \alpha \mathbf{B}, \tag{2.87}$$

where **M** and **P** denote the changes of magnetization and polarization from those in the system without electric field and magnetic field, respectively. From the Maxwell equation, to generate the magnetic (electric) field by the electric (magnetic) field, the time variation of the electric (magnetic) field is necessary. Therefore, the magnetoelectric effect is not allowed in non-magnetic crystals because this effect violates time-reversal symmetry. In order to realize the magnetoelectric effect, it is necessary to break the time-reversal and inversion symmetries. In fact, the magnetoelectric effect can be realized in a magnetic crystal with nontrivial magnetic structure.

The magnetoelectric effect was theoretically predicted first by Curie, and later, Dzyaloshinskii predicted the linear magnetoelectric effect in $Cr_2O_3$ by using the crystal symmetry argument [33]. Then, Astrov [34] and Folen et al. [35] observed experimentally the variation of magnetization with the electric field in a single crystal of $Cr_2O_3$. The variation in its magnetization in response to an alternating electric field has been detected, and it has been shown that the effect is linear in the applied electric field. The form of the tensor $\alpha$ in $Cr_2O_3$ is consistent with the form determined from the magnetic point group $\bar{3}'m' = 32 + \mathrm{TR} \times (\bar{3}m - 32)$. We note that the prime means a joint operation of a spatial operation and the time-reversal operation TR. This effect is also observed in other crystals such as $GaFeO_3$ [36].

## References

1. Zhang L, Niu Q (2014) Angular momentum of phonons and the einstein-de haas effect. Phys Rev Lett 112:085503
2. Zhang L, Niu Q (2015) Chiral phonons at high-symmetry points in monolayer hexagonal lattices. Phys Rev Lett 115:115502
3. Zhu H et al (2018) Observation of chiral phonons. Science 359:579–582
4. Srivastava G (1990) The physics of phonons. Taylor & Francis
5. Kittel C (2004) Introduction to solid state physics, 8th edn. Wiley
6. Ashcroft NW, Mermin ND (1976) Solid state physics. Holt-Saunders
7. Grosso G, Parravicini G (2000) Solid state physics. Elsevier Science
8. Sheng L, Sheng DN, Ting CS (2006) Theory of the phonon hall effect in paramagnetic dielectrics. Phys Rev Lett 96:155901
9. Liu Y, Xu Y, Zhang S-C, Duan W (2017) Model for topological phononics and phonon diode. Phys Rev B 96:064106

10. Wang J-S, Zhang L (2009) Phonon hall thermal conductivity from the green-kubo formula. Phys Rev B 80:012301
11. Zhang L, Ren J, Wang J-S, Li B (2010) Topological nature of the phonon hall effect. Phys Rev Lett 105:225901
12. Kagan Y, Maksimov LA (2008) Anomalous hall effect for the phonon heat conductivity in paramagnetic dielectrics. Phys Rev Lett 100:145902
13. Zhang L, Wang J-S, Li B (2009) Phonon hall effect in four-terminal nano-junctions. New J Phys 11:113038
14. Zhang L, Ren J, Wang J-S, Li B (2011) The phonon hall effect: theory and application. J Phys Condens Matter 23:305402
15. Holz A (1972) Phonons in a strong static magnetic field. Il Nuovo Cimento B 1971–1996(9):83–95
16. de L Kronig R (1939) On the mechanism of paramagnetic relaxation. Physica 6:33–43
17. Van Vleck JH (1940) Paramagnetic relaxation times for titanium and chrome alum. Phys Rev 57:426–447
18. Capellmann H, Lipinski S (1991) Spin-phonon coupling in intermediate valency: exactly solvable models. Zeitschrift für Physik B Condens Matter 83:199–205
19. Capellmann H, Neumann KU (1987) Intermediate valency i: couplings, hamiltonian, and electrical resistivity. Zeitschrift für Physik B Condens Matter 67:53–61
20. Capellmann H, Lipinski S, Neumann K-U (1989) A microscopic model for the coupling of spin fluctuations and charge fluctuation in intermediate valency. Zeitschrift fur Physik B Condens Matter 75:323–329
21. Wang Y-T, Luan P-G, Zhang S (2015) Coriolis force induced topological order for classical mechanical vibrations. New J Phys 17:073031
22. Wang P, Lu L, Bertoldi K (2015) Topological phononic crystals with one-way elastic edge waves. Phys Rev Lett 115:104302
23. Nash LM et al (2015) Topological mechanics of gyroscopic metamaterials. Proc Natl Acad Sci 112:14495–14500
24. Chen S et al (2012) Thermal conductivity of isotopically modified graphene 11:203–207
25. Kim G et al (2013) Growth of high-crystalline, single-layer hexagonal boron nitride on recyclable platinum foil. Nano Lett 13:1834–1839
26. Birss RR (1962) In: Wohlfarth EP (ed) Symmetry and magnetism, vol 3. Series of monographs on selected topics in solid state physics. Elsevier North-Holland
27. Bradley C, Cracknell A (1972) The mathematical theory of symmetry in solids: representation theory for point groups and space groups. Clarendon Press
28. Burns G (1977) Introduction to group theory with applications. Materials science and technology. Academic Press
29. Edelstein VM (1990) Spin polarization of conduction electrons induced by electric current in two-dimensional asymmetric electron systems. Solid State Commun 73:233–235
30. Ivchenko EL, Pikus GE (1978) New photogalvanic effect in gyrotropic crystals 27:604
31. Levitov LS, Nazarov YV, Eliashberg GM (1985) Magnetoelectric effects in conductors with mirror isomer symmetry 61:133
32. Aronov AG, Lyanda-Geller YB (1989) Nucleasr electric resonance and orientation of carier spins by an electric field 50:431
33. Dzyaloshinskii IE (1959) On the magneto-electrical effect in antiferromagnets 10:628
34. Astrov DN (1960) The magnetoelectric effect in antiferromagnetics 11:728
35. Folen VJ, Rado GT, Stalder EW (1961) Anisotropy of the magnetoelectric effect in $cr_2o_3$. Phys Rev Lett 6:607–608
36. Arima T et al (2004) Structural and magnetoelectric properties of $ga_{2-x}fe_xo_3$ single crystals grown by a floating-zone method. Phys Rev B 70:064426

# Chapter 3
# Phonon Thermal Edelstein Effect

## 3.1 Phonon Thermal Edelstein Effect

Phonon angular momentum can be nonzero in the system without inversion or time-reversal symmetry. In previous theory [1], the phonon angular momentum becomes nonzero in the system without time-reversal symmetry. In this chapter, we focus on the phonon in the sysmte without inversion symmetry. When the system has time-reversal symmetry and does not have inversion symmetry, the phonon angular momentum of each phonon mode $\mathbf{l}_\sigma(\mathbf{k})$ is an odd function of wave vector $\mathbf{k}$. The total phonon angular momentum $\mathbf{J}^{\rm ph}$ becomes zero, because the phonon angular momentum of each phonon mode cancels between $\mathbf{k}$ and $-\mathbf{k}$, as follows

$$J_i^{\rm ph} = \sum_{\mathbf{k},\sigma} l_{\sigma,i}(\mathbf{k})\left(f_0(\omega_\sigma(\mathbf{k})) + \frac{1}{2}\right) \tag{3.1}$$

$$\begin{aligned} &= \sum_{\mathbf{k}\geq 0,\sigma}\left[l_{\sigma,i}(\mathbf{k})\left(f_0(\omega_\sigma(\mathbf{k})) + \frac{1}{2}\right) + l_{\sigma,i}(-\mathbf{k})\left(f_0(\omega_\sigma(-\mathbf{k})) + \frac{1}{2}\right)\right] \\ &= \sum_{\mathbf{k}\geq 0,\sigma}\left[(l_{\sigma,i}(\mathbf{k}) + l_{\sigma,i}(-\mathbf{k}))\left(f_0(\omega_\sigma(\mathbf{k})) + \frac{1}{2}\right)\right] \\ &= 0, \end{aligned} \tag{3.2}$$

where $f_0(\omega_\sigma(\mathbf{k})) = (\exp(\hbar\omega_\sigma(\mathbf{k})/k_{\rm B}T) - 1)^{-1}$ is the Bose distribution function, $k_{\rm B}$ is the Boltzmann constant, $T$ is the temperature, $\mathbf{k} = (k_x, k_y, k_z)$ is the wavevector, and $i = x, y, z$ is a real-space coordinate. Here we used the relations: $\mathbf{l}_\sigma(\mathbf{k}) = -\mathbf{l}_\sigma(-\mathbf{k})$, $\omega_\sigma(\mathbf{k}) = \omega_\sigma(-\mathbf{k})$. On the other hand, we will show that when the temperature gradient is applied, the phonon angular momentum becomes nonzero. We introduce the Boltzmann transport theory of phonons [2, 3]. In thermal equilibrium, the phonons are distributed according to the Bose distribution function, and the phonon distribution $f_0$ does not depend on the position $\mathbf{x}$ and the time $t$. In systems with a temperature gradient $\partial T/\partial x_i$, the phonon distribution $f_{\sigma,\mathbf{k}}(\mathbf{x}, t)$ depends on



M. Hamada, *Theory of Generation and Conversion of Phonon Angular Momentum*, Springer Theses, https://doi.org/10.1007/978-981-33-4690-1_3

the position $\mathbf{x}$ and the time $t$. The phonon distribution changes with time in two ways; one is a diffusion from point to point, and another is a scattering due to nonlinear phonon-phonon interactions. The phonon distribution $f_{\sigma,\mathbf{k}}(\mathbf{x}, t)$ is represented as

$$\frac{df_{\sigma,\mathbf{k}}}{dt} = \left.\frac{\partial f_{\sigma,\mathbf{k}}}{\partial t}\right|_{\text{diff.}} + \left.\frac{\partial f_{\sigma,\mathbf{k}}}{\partial t}\right|_{\text{scatt.}}. \tag{3.3}$$

Here, we assume that the temperature does not depend on time, which means the heat current is in a steady state. Thus, the total derivative with respect to time $df_{\sigma,\mathbf{k}}/dt$ is zero. The scattering term is written as

$$\left.\frac{\partial f_{\sigma,\mathbf{k}}}{\partial t}\right|_{\text{scatt.}} = -\frac{f_{\sigma,\mathbf{k}} - f_0}{\tau}, \tag{3.4}$$

with the phonon relaxation time $\tau$. The diffusion term is related to the temperature gradient $\partial T/\partial x_i$, and this term is represented as

$$\begin{aligned}\left.\frac{\partial f_{\sigma,\mathbf{k}}}{\partial t}\right|_{\text{diff.}} &= \lim_{dt\to 0}\frac{f_{\sigma,\mathbf{k}}(\mathbf{x} - v_{\sigma,i}(\mathbf{k})dt) - f_{\sigma,\mathbf{k}}(\mathbf{x})}{dt}\\ &= -v_{\sigma,i}(\mathbf{k})\frac{\partial f_{\sigma,\mathbf{k}}}{\partial x_i} = -v_{\sigma,i}(\mathbf{k})\frac{\partial f_0}{\partial T}\frac{\partial T}{\partial x_i},\end{aligned} \tag{3.5}$$

where $v_{\sigma,i}(\mathbf{k}) = \partial\omega_\sigma(\mathbf{k})/\partial k_i$ is the group velocity of each mode. In the third equality, we replace $f_{\sigma,\mathbf{k}}$ by $f_0$ since we assume steady state and the local thermal equilibrium. By substituting Eqs. (3.4) and (3.5) into Eq. (3.3), the phonon distribution $f_{\sigma,\mathbf{k}}$ is represented as

$$f_{\sigma,\mathbf{k}} = f_0(\omega_\sigma(\mathbf{k})) - \tau v_{\sigma,i}(\mathbf{k})\frac{\partial f_0}{\partial T}\frac{\partial T}{\partial x_i}. \tag{3.6}$$

To justify the use of the Boltzmann transport theory, we assume here that the deviation of the system away from equilibrium is small. In order to satisfy this condition, we focus on the linear response regime where the heat current is infinitesimally small. We also assume that the system relaxes towards the local thermal equilibrium quickly via nonlinear phonon-phonon interactions. As shown in Eq. (3.6), the phonon relaxation time $\tau$ based on the constant relaxation time approximation represents the effect of nonlinear phonon-phonon interactions. The dependence of $\tau$ on the mode index $\sigma$ and the wavevector $\mathbf{k}$ does not change our main conclusion and the constant relaxation time approximation is enough for a rough estimation. By substituting Eq. (3.6) into Eq. (3.1), the total phonon angular momentum per unit volume becomes

$$J_i^{\text{ph}} = -\frac{\tau}{V}\sum_{\mathbf{k},\sigma} l_{\sigma,i}(\mathbf{k})v_{\sigma,j}(\mathbf{k})\frac{\partial f_0(\omega_\sigma(\mathbf{k}))}{\partial T}\frac{\partial T}{\partial x_j} \equiv \alpha_{ij}\frac{\partial T}{\partial x_j} \tag{3.7}$$

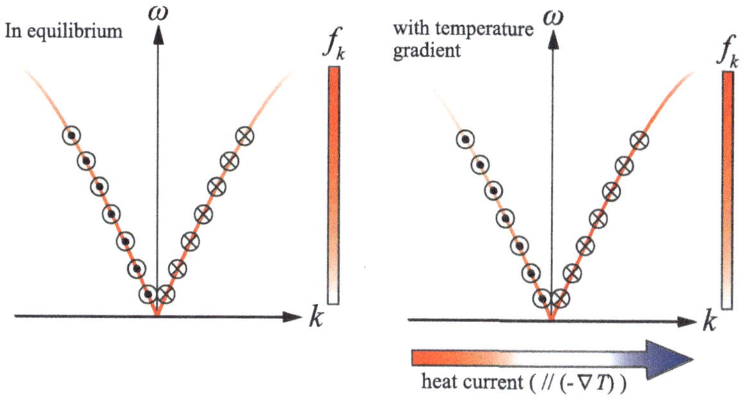


**Fig. 3.1** Schematic figures of the phonon dispersion in equilibrium (left panel) and that with temperature gradient (right panel). The colormap shows the distribution of phonons following the distribution function $f_{\mathbf{k}}$. The circles with filled circles and crosses show the directions of phonon angular momentum, which are taken to be perpendicular to $k$. Reprinted figure with permission from [11] Masato Hamada, Emi Minamitani, Motoaki Hirayama, and Shuichi Murakami, "Phonon Angular Momentum Induced by the Temperature Gradient", Phys. Rev. Lett. 121. 175301, 2018, Copyright 2018 by the American Physical Society

where $\alpha_{ij}$ denotes a response tensor. The generated phonon angular momentum is proportional to the temperature gradient. This effect is caused by nonequilibrium phonon distribution, leading to an unbalance of phonon angular momentum as shown in Fig. 3.1. Since this is analogous to the Edelstein effect [4–10] in electronic systems, we call this effect phonon thermal Edelstein effect [11].

## 3.2 Symmetry Constraint for the Phonon Edelstein Effect

In order to realize a nonzero response tensor $\alpha_{ij}$, the crystal symmetry should be sufficiently low. Because it is a response at $\mathbf{k} = 0$, the nonzero elements of the tensor are determined only by the point-group symmetry, and not by the space-group symmetry. This response tensor $\alpha_{ij}$ is an axial tensor, and then one can easily identify nonzero elements of the tensor. First, among 32 point groups, the all elements of the axial tensor becomes zero in the 11 point groups having inversion symmetry. Among the remaining 21 point groups without inversion symmetry, the axial tensor $\alpha_{ij}$ has nonzero elements in 18 point groups: $O$, $T$, $D_4$, $D_{2d}$, $C_{4v}$, $S_4$, $C_4$, $D_2$, $C_{2v}$, $D_6$, $C_{6v}$, $C_6$, $D_3$, $C_{3v}$, $C_3$, $C_{1h}$, $C_2$, and $C_1$. For example, we show the axial tensor $\alpha_{ij}$ in systems with the point group $C_{6v}$. The generators of the point group $C_{6v}$ are represented as

$$C_6 = \begin{pmatrix} \frac{1}{2} & -\frac{\sqrt{3}}{2} & 0 \\ \frac{\sqrt{3}}{2} & \frac{1}{2} & 0 \\ 0 & 0 & 1 \end{pmatrix}, \quad \sigma_x = \begin{pmatrix} -1 & 0 & 0 \\ 0 & 1 & 0 \\ 0 & 0 & 1 \end{pmatrix}, \tag{3.8}$$

where $C_6$ is the six-fold rotational operator around the $z$ axis, and $\sigma_x$ is the mirror operation with respect to the $yz$ plane. Because the axial tensor $\alpha$ should satisfy $C_6 \alpha C_6^{-1} = \alpha$ and $-\sigma_x \alpha \sigma_x^{-1} = \alpha$, where $\alpha$ is the matrix consisting of the elements $\alpha_{ij}$, the tensor is represented as

$$\alpha = \begin{pmatrix} 0 & \alpha_{xy} & 0 \\ -\alpha_{xy} & 0 & 0 \\ 0 & 0 & 0 \end{pmatrix}. \tag{3.9}$$

In fact, a full list of the form of the axial tensor for the 18 point groups is available from [12], as follows:

$$\alpha = \begin{pmatrix} \alpha_{xx} & 0 & 0 \\ 0 & \alpha_{xx} & 0 \\ 0 & 0 & \alpha_{xx} \end{pmatrix} : \ O,\ T, \tag{3.10}$$

$$\alpha = \begin{pmatrix} \alpha_{xx} & 0 & 0 \\ 0 & \alpha_{xx} & 0 \\ 0 & 0 & \alpha_{zz} \end{pmatrix} : \ D_4,\ D_6,\ D_3, \tag{3.11}$$

$$\alpha = \begin{pmatrix} \alpha_{xx} & \alpha_{xy} & 0 \\ -\alpha_{xy} & \alpha_{xx} & 0 \\ 0 & 0 & \alpha_{zz} \end{pmatrix} : \ C_4,\ C_6,\ C_3, \tag{3.12}$$

$$\alpha = \begin{pmatrix} \alpha_{xx} & \alpha_{xy} & 0 \\ \alpha_{xy} & -\alpha_{xx} & 0 \\ 0 & 0 & 0 \end{pmatrix} : \ S_4, \tag{3.13}$$

$$\alpha = \begin{pmatrix} 0 & \alpha_{xy} & 0 \\ -\alpha_{xy} & 0 & 0 \\ 0 & 0 & 0 \end{pmatrix} : \ C_{4v},\ C_{6v},\ C_{3v}, \tag{3.14}$$

$$\alpha = \begin{pmatrix} \alpha_{xx} & 0 & 0 \\ 0 & -\alpha_{xx} & 0 \\ 0 & 0 & 0 \end{pmatrix} : \ D_{2d}, \tag{3.15}$$

$$\alpha = \begin{pmatrix} \alpha_{xx} & 0 & 0 \\ 0 & \alpha_{yy} & 0 \\ 0 & 0 & \alpha_{zz} \end{pmatrix} : \ D_2, \tag{3.16}$$

$$\alpha = \begin{pmatrix} 0 & \alpha_{xy} & 0 \\ \alpha_{yx} & 0 & 0 \\ 0 & 0 & 0 \end{pmatrix} : \ C_{2v}, \tag{3.17}$$

$$\alpha = \begin{pmatrix} 0 & 0 & \alpha_{xz} \\ 0 & 0 & \alpha_{yz} \\ \alpha_{zx} & \alpha_{zy} & 0 \end{pmatrix} : \ C_{1h}, \tag{3.18}$$

$$\alpha = \begin{pmatrix} \alpha_{xx} & \alpha_{xy} & 0 \\ \alpha_{yx} & \alpha_{yy} & 0 \\ 0 & 0 & \alpha_{zz} \end{pmatrix} : \ C_2, \tag{3.19}$$

$$\alpha = \begin{pmatrix} \alpha_{xx} & \alpha_{xy} & \alpha_{xz} \\ \alpha_{yx} & \alpha_{yy} & \alpha_{yz} \\ \alpha_{zx} & \alpha_{zy} & \alpha_{zz} \end{pmatrix} : \ C_1. \tag{3.20}$$

We note that symmetry constraints described in Eqs. (3.10)–(3.20) are the same as the Edelstein effect in electronic systems. It is instructive to decompose the response tensor into symmetric and antisymmetric parts. The antisymmetric part of $\alpha_{ij}$ is essentially a polar vector $\alpha_k \equiv \epsilon_{ijk}\alpha_{ij}$ and therefore it survives only for polar crystals, such as polar metals and ferroelectrics. In this case, when we set the $z$ axis to be along the polarization or the polar axis, $\alpha_{xy} = -\alpha_{yx}$ are the only nonzero elements of this tensor. Thus, the temperature gradient and the generated angular momentum are perpendicular to each other, and they are both perpendicular to the polarization vector in any polar crystals. On the other hand, the symmetric part of $\alpha_{ij}$ changes sign under inversion, and remains typically in chiral systems such as tellurium and selenium. Both the antisymmetric and the symmetric parts become nonzero in systems with very low symmetry.

## 3.3 Generation of Phonon Angular Momentum by Temperature Gradient

In this section, we calculate the phonon angular momentum generated by the temperature gradient [11]. We focus on a polar crystal and a chiral crystal, and we calculate the phonon angular momentum by using the first-principle calculations.

### 3.3.1 *Polar Crystal: Wurtzite Gallium Nitride*

In this section, as an example of polar crystals, we discuss the wurtzite structure (space group: $P6_3mc$). The wurtzite structure has four atoms in the unit cell, as shown in Fig. 3.2a for gallium nitride (GaN). When we take the polar axis to be along the $z$ axis, the nonzero elements of the response tensor $\alpha_{ij}$ are $\alpha_{xy} = -\alpha_{yx}$ from symmetry analysis, as shown in Fig. 3.2c. We first calculate the phonon angular momentum generated by heat current both by using (i) the valence force field model of Keating [13] and by using (ii) first-principle calculations. We show the numerical results

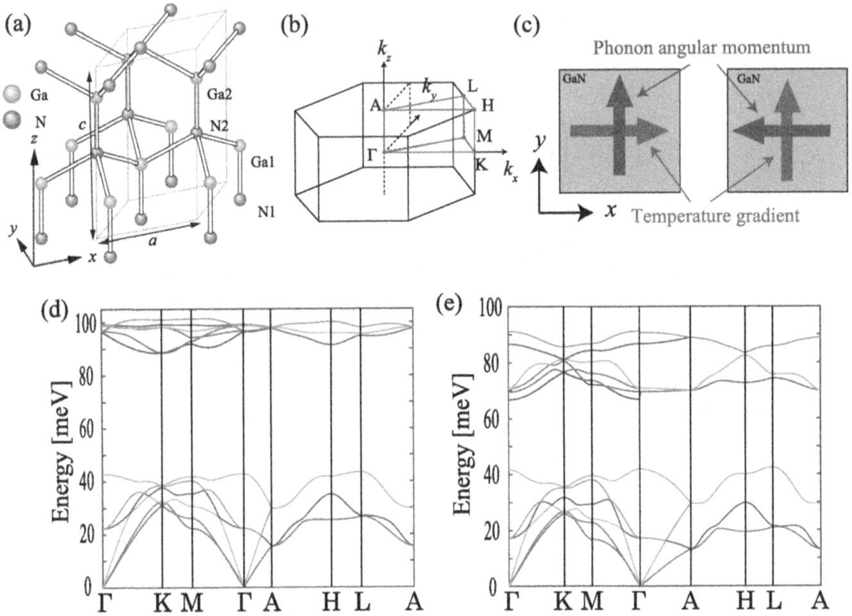


**Fig. 3.2** Crystal structure of wurtzite GaN and the phonon angular momentum of wurtzite GaN. **a** Crystal structure of wurtzite GaN and the red line shows the unit cell. **b** First Brillouin zone of wurtzite GaN. **c** Schematic illustration of the relation between temperature gradient and the phonon angular momentum. **d** Numerical results of phonon dispersion of wurtzite GaN using the valence force field model. **e** Numerical results of phonon dispersion of wurtzite GaN using the first-principle calculation. Reprinted figure with permission from [11] Masato Hamada, Emi Minamitani, Motoaki Hirayama, and Shuichi Murakami, "Phonon Angular Momentum Induced by the Temperature Gradient", Phys. Rev. Lett. 121. 175301, 2018, Copyright 2018 by the American Physical Society

of phonon dispersion for the valence force field model and for the first-principle calculation in Fig. 3.2d, e.

### (i) Valence Force Field Model of Keating for Wurtzite GaN

We describe the details of the valence force field model. Here, we use the parameters of GaN for calculations of the phonon angular momentum [13]. For simplicity, in this model we take the inter-atomic forces up to the next nearest neighbors. The elastic energy of the valence force field model is represented as

$$
\begin{aligned}
U_i = & \frac{3\alpha}{16r_0^2}\sum_{j=1}^{3}(\mathbf{r}_{ij}^2 - r_0^2)^2 + \frac{3\alpha'}{16r_0'^2}(\mathbf{r}_{i4}^2 - r_0'^2)^2 \\
& + \frac{3\beta}{8r_0^2}\sum_{j=1}^{3}\sum_{k>j}^{3}(\mathbf{r}_{ij}\cdot\mathbf{r}_{ik} - r_0^2\cos\theta_0)^2 \\
& + \frac{3\beta'}{8r_0'^2}\sum_{k=1}^{3}(\mathbf{r}_{i4}\cdot\mathbf{r}_{ik} - r_0^2\cos\theta_0')^2
\end{aligned} \tag{3.21}
$$

where $r_0$ is the equilibrium length between the Ga1 atom and the neighboring N2 atom, $r_0'$ is the equilibrium length along $c$-axis between the Ga1(Ga2) atom and the N1(N2) atom, and $\theta_0$ and $\theta_0'$ are bond angle Ga1-N2-Ga1 and bond angle Ga1-N2-Ga2. The lattice constants are $a = 3.189$ Å, $c = 5.185$ Å and the lattice parameter is $u = 0.3768$, which gives an equilibrium value of $r_0'/c$. The four parameters $\alpha$, $\beta$, $\alpha'$, and $\beta'$ are Keating parameters, with $\alpha, \alpha'$ being bond-stretching constants and $\beta, \beta'$ being bond-bending constants. We adopt the Keating parameters as $\alpha = \alpha' = 88.35$ N/m and $\beta = \beta' = 20.92$ N/m. We calculate the inter-atomic force constants from this Keating potential, and calculate the dynamical matrix $D$. Here, we ignore the effect of polarization in wurtzite GaN in this model for simplicity of the calculation.

We show the distribution of phonon angular momentum of GaN by the valence force field model. The phonon dispersion is obtained as Fig. 3.2d with the Brillouin zone in Fig 3.2b. Nevertheless, despite its simplicity, this model well describes the nature of phonon angular momentum. The phonon bands below the frequency gap are in good agreement with first-principle calculation, but the phonon bands above the frequency gap are not. Nevertheless, this model is useful for extracting the distribution of phonon angular momentum in a semi-quantitative manner. In Fig. 3.3a and c, we show the distributions of the angular momentum of phonons $l_\sigma(\mathbf{k})$of the third and sixth lowest bands on the plane $k_z = 0$. As illustrative examples, Fig. 3.3b, d show the trajectories of the four atoms in the unit cell for particular choices of the mode $\sigma$ and the wavenumber $\mathbf{k}$ shown as black dots in Fig. 3.3a, c, respectively.

### (ii) First-Principle Calculations for GaN

We describe the details of our first-principle calculation. The phonon properties in wurtzite GaN are calculated by using the density functional perturbation theory (DFPT) [14] implemented in the Quantum-Espresso package [15]. We use the ultrasoft pseudopotentials generated by Garrity-Bennett-Rabe-Vanderbilt (GBRV) method [16] for Ga and N atoms. The local-density approximation is adopted for the exchange-correlation functional in GaN. The expansion of the plane-wave set is restricted by a kinetic energy cutoff of 80 Ry for GaN. The optimized lattice constants in GaN are $a = 3.155$ Å and $c = 5.143$ Å. A $12 \times 12 \times 12$ ($6 \times 6 \times 6$) Monkhorst-Pack grid [17] without an offset is used for self-consistent electronic

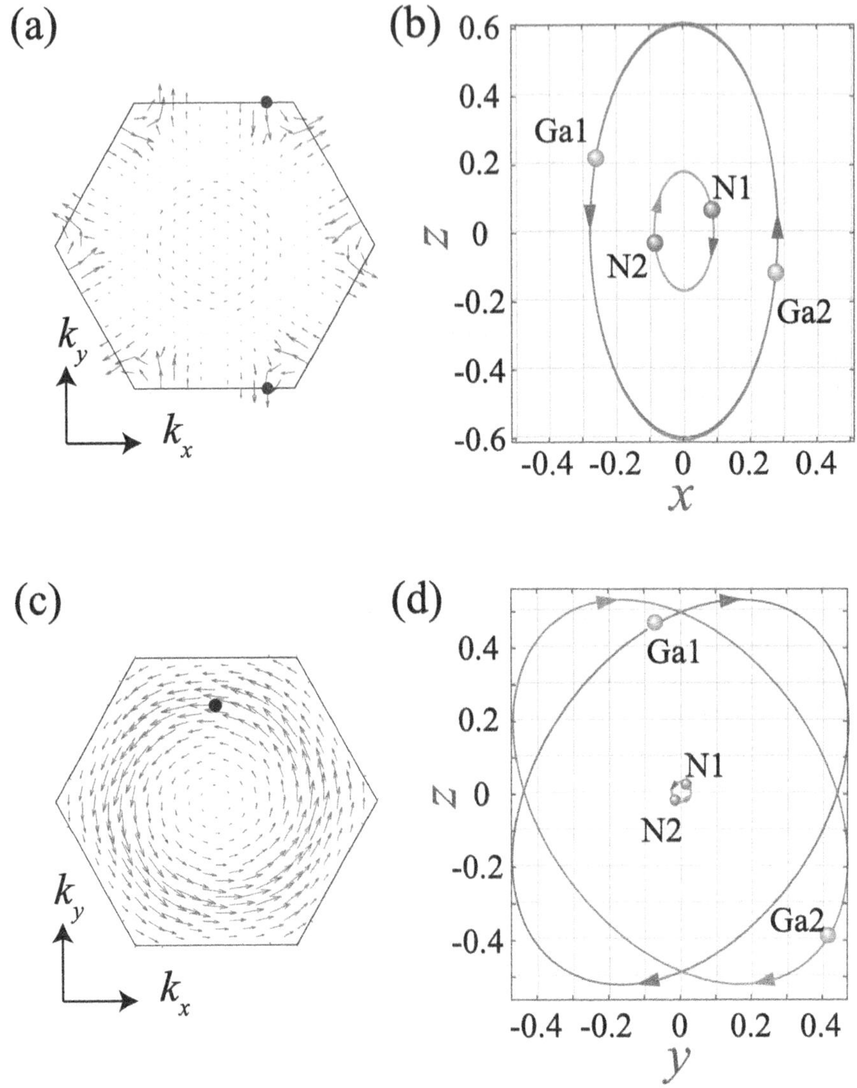


**Fig. 3.3** **a** Distribution of the angular momentum of phonons $l_\sigma(\mathbf{k})$ of the third lowest band on the plane $k_z = 0$. **b** Trajectories of the four atoms in the unit cell for the phonon mode of the third band at $\frac{\mathrm{k}a}{2\pi} = (0.2222, \frac{1}{\sqrt{3}}, 0)$, which is indicated as a black dot in **a**. **c** Distribution of the angular momentum of phonons $l_\sigma(\mathbf{k})$ of the sixth lowest band on the plane $k_z = 0$. **d** Trajectories of the four atoms in the unit cell for the phonon mode of the sixth band at $\frac{\mathrm{k}a}{2\pi} = (0, 0.3849, 0)$, which is indicated as a black dot in **c**. **b** and **d** represent the normalized polarization vector $\varepsilon_\sigma(\mathbf{k})$ with $\varepsilon^\dagger\varepsilon = 1$, and their axes $(x, y, z)$ are shown in a dimensionless unit. Reprinted figure with permission from [11] Masato Hamada, Emi Minamitani, Motoaki Hirayama, and Shuichi Murakami, "Phonon Angular Momentum Induced by the Temperature Gradient", Phys. Rev. Lett. 121. 175301, 2018, Copyright 2018 by the American Physical Society

structure calculations (phonon calculations) for GaN. The nonanalytic term of the macroscopic electric field induced by the optical phonons is also included in wurtzite GaN calculation.

We show the distribution of the phonon angular momentum of GaN by the first-principle calculation. The phonon dispersion is obtained as Fig. 3.2e with the Bril-

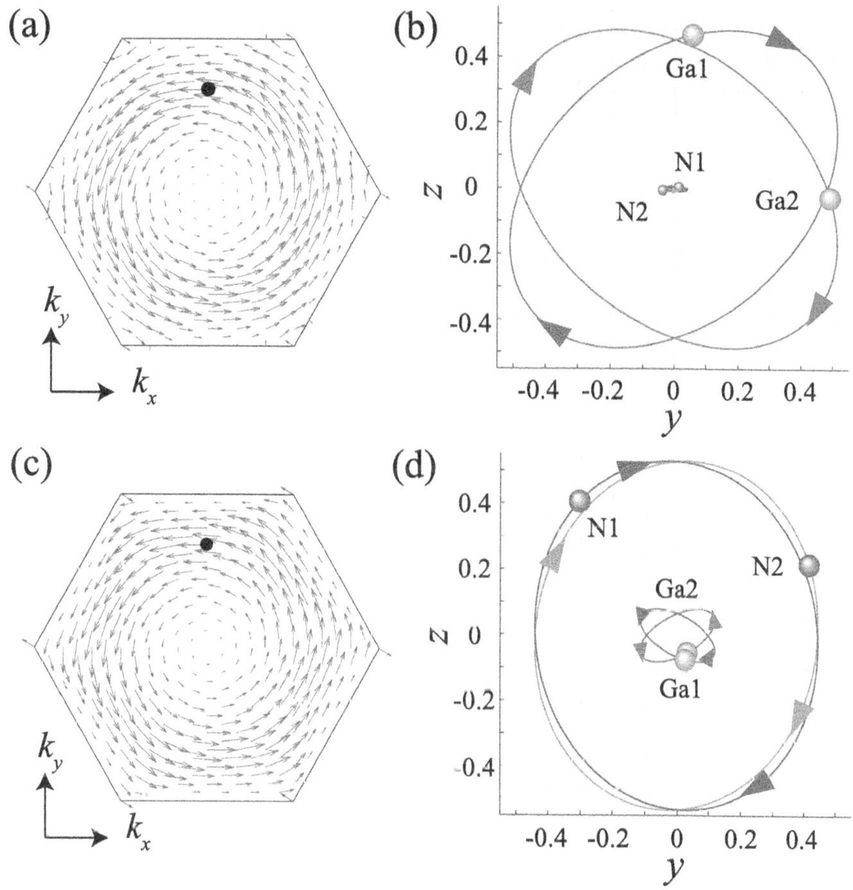


**Fig. 3.4** **a** Distribution of the phonon angular momentum $\mathbf{l}_\sigma(\mathbf{k})$ of the sixth lowest band on the plane $k_z = 0$. **b** Trajectories of the four atoms in the unit cell for the phonon of the sixth band at $\frac{\mathbf{k}a}{2\pi} = (0, 0.3849, 0)$, which is indicated as a black dot in **a**. **c** Distribution of the phonon angular momentum $\mathbf{l}_\sigma(\mathbf{k})$ of the eleventh lowest band on the plane $k_z = 0$. **d** Trajectories of the four atoms in the unit cell for the phonon of the eleventh band at $\frac{\mathbf{k}a}{2\pi} = (0, 0.3849, 0)$, which is indicated as a black dot in **c**. **b** and **d** represent the normalized polarization vector $\varepsilon_\sigma(\mathbf{k})$ with $\varepsilon^\dagger\varepsilon = 1$, and their axes $(x, y, z)$ are shown in a dimensionless unit. Reprinted figure with permission from [11] Masato Hamada, Emi Minamitani, Motoaki Hirayama, and Shuichi Murakami, "Phonon Angular Momentum Induced by the Temperature Gradient", Phys. Rev. Lett. 121. 175301, 2018, Copyright 2018 by the American Physical Society

**Table 3.1** Numerical results of induced phonon angular momentum and a rigid-body rotation of GaN. Modified table with permission from [11] Masato Hamada, Emi Minamitani, Motoaki Hirayama, and Shuichi Murakami, "Phonon Angular Momentum Induced by the Temperature Gradient", Phys. Rev. Lett. 121. 175301, 2018, Copyright 2018 by the American Physical Society

| | Valence force field model | First-principle |
|---|---|---|
| $a$ [Å] | 3.189 | 3.155 |
| $c$ [Å] | 5.185 | 5.143 |
| $\rho$ [kg/m$^3$] | $6.089 \times 10^3$ | $6.276 \times 10^3$ |
| $\alpha_{xy}$ below the frequency gap [Jsm$^{-2}$K$^{-1}$] | $-7.4 \times 10^{-8} \times \tau$ | $-8.9 \times 10^{-8} \times \tau$ |
| $\alpha_{xy}$ above the frequency gap [Jsm$^{-2}$K$^{-1}$] | $1.8 \times 10^{-7} \times \tau$ | $-2.7 \times 10^{-7} \times \tau$ |
| total $\alpha_{xy}$ [Jsm$^{-2}$K$^{-1}$] | $1.1 \times 10^{-7} \times \tau$ | $-3.6 \times 10^{-7} \times \tau$ |
| $\alpha_{xy}^{\mathrm{Ga}}$ [Jsm$^{-2}$K$^{-1}$] | – | $-7.1 \times 10^{-8} \times \tau$ |
| $\alpha_{xy}^{\mathrm{N}}$ [Jsm$^{-2}$K$^{-1}$] | – | $-2.9 \times 10^{-7} \times \tau$ |
| $\Omega_{\mathrm{rigid-body}}$ [s$^{-1}$] | $-1.1 \times 10^{-21} \times \Delta T/L^3$ | $3.4 \times 10^{-21} \times \Delta T/L^3$ |

louin zone in Fig. 3.2b. The band structure obtained by the first-principle calculation (Fig. 3.2e) shows good agreement with previous works [18–20]. Overall features of the band structure are similar to those obtained by the valence force field model except for the splitting of the longitudinal and transverse optical bands at the long wavelength limit. Examples of distributions of phonon angular momentum $\mathbf{l}_\sigma(\mathbf{k})$ are in Fig. 3.4a and c, showing similarity with spin structure in Rashba systems. We show the trajectories of atoms in the sixth and eleventh lowest modes in Fig. 3.4b and d, respectively. By comparing Fig. 3.4a and c, the oscillation of nitrogen atoms in the eleventh modes is much larger than that of the gallium atoms, while the oscillation of gallium atoms in the sixth mode is larger. The response tensor is estimated as $\alpha_{xy} \sim -10^{-7} \times [\tau/(1\mathrm{s})]\,\mathrm{Jsm}^{-2}\mathrm{K}^{-1}$ at $T = 300$ K. We show the numerical results by using the valence force field model and by the first-principle calculation in Table 3.1.

### 3.3.2 *Chiral Crystals: Tellurium and Selenium*

As other examples of chiral crystals, we consider Te (tellurium) and Se (selenium) [21]. Te and Se have a helical crystal structure, as shown in Fig. 3.5a. The helical chains having three atoms in a unit cell form a triangular lattice. The space group is $P3_121$ or $P3_221$ ($D_3^4$ or $D_3^6$) corresponding to the right-handed or left-handed screw symmetry. They are semiconductors at ambient pressure. We describe the details of our first-principle calculations for Te and Se. The phonon properties in Te and Se crystals are calculated by using the density functional perturbation theory (DFPT) [14] implemented in the Quantum-Espresso package [15]. We use the ultrasoft pseudopotentials generated by Garrity-Bennett-Rabe-Vanderbilt (GBRV) method [16] for Te,

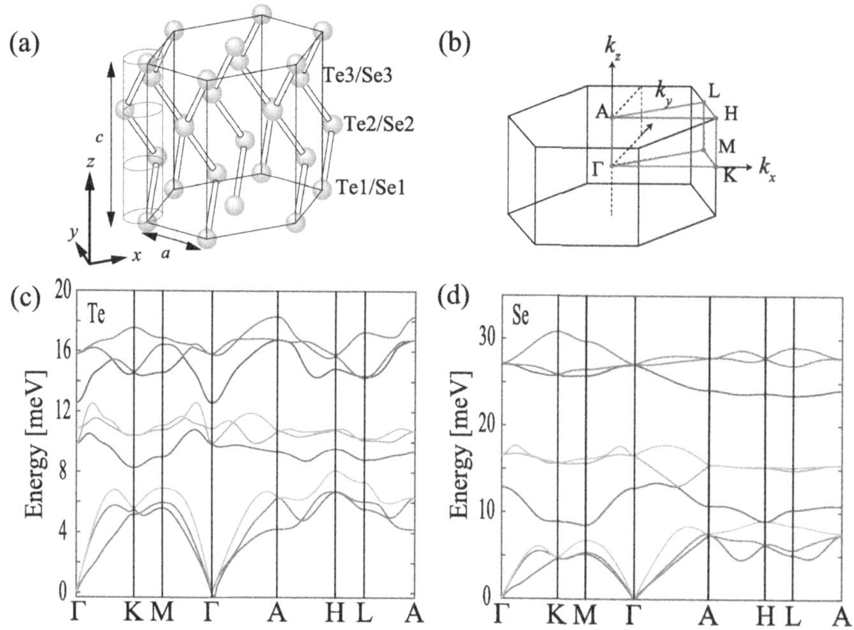


**Fig. 3.5** **a** Crystal structure of Te and Se. **b** First Brillouin zone of Te and Se. **c** Phonon dispersion of Te. **d** Phonon dispersion of Se. Reprinted figure with permission from [11] Masato Hamada, Emi Minamitani, Motoaki Hirayama, and Shuichi Murakami, "Phonon Angular Momentum Induced by the Temperature Gradient", Phys. Rev. Lett. 121. 175301, 2018, Copyright 2018 by the American Physical Society

and Se atoms. The generalized gradient approximation is adopted in Te and Se. The expansion of the plane-wave set is restricted by a kinetic energy cutoff of 30 Ry for Te and Se, respectively. The optimized intra-atomic/inter-atomic bond-length is $r = 2.373$ Å and $R = 3.440$ Å in Se and $r = 2.834$ Å and $R = 3.491$ Å in Te, respectively. A $12 \times 12 \times 8$ ($8 \times 8 \times 8$) Monkhorst-Pack grid [17] without an offset is used for self-consistent electronic structure calculations (phonon calculations) for Se and Te. Numerical results of the phonon dispersions of Te and Se by first-principle calculation are shown in Fig. 3.5c and d with first Brillouin zone Fig 3.5b, respectively. The distributions of phonon angular momentum $\mathbf{l}_\sigma(\mathbf{k})$ on the two planes in the Brillouin zone (Fig. 3.6a) are shown in Fig. 3.6b, c for the fourth lowest band. Because of the threefold screw symmetry around the $z$ axis, the angular momentum on the $k_z$ axis is along the $z$ axis. Figure 3.6d represents the trajectories of the three atoms in the unit cell for the fourth lowest band. Here, because of the threefold screw symmetry at the **k** point considered, the trajectories are related with each other by threefold rotation around the $z$ axis, and the angular momentum is along the $z$ axis by symmetry. In Te and Se, from symmetry argument, the response tensor has nonzero elements $\alpha_{xx} = \alpha_{yy}$ and $\alpha_{zz}$, whose symmetry is identical with the electronic Edelstein effect in tellurium [10]. The response tensor for Te is estimated

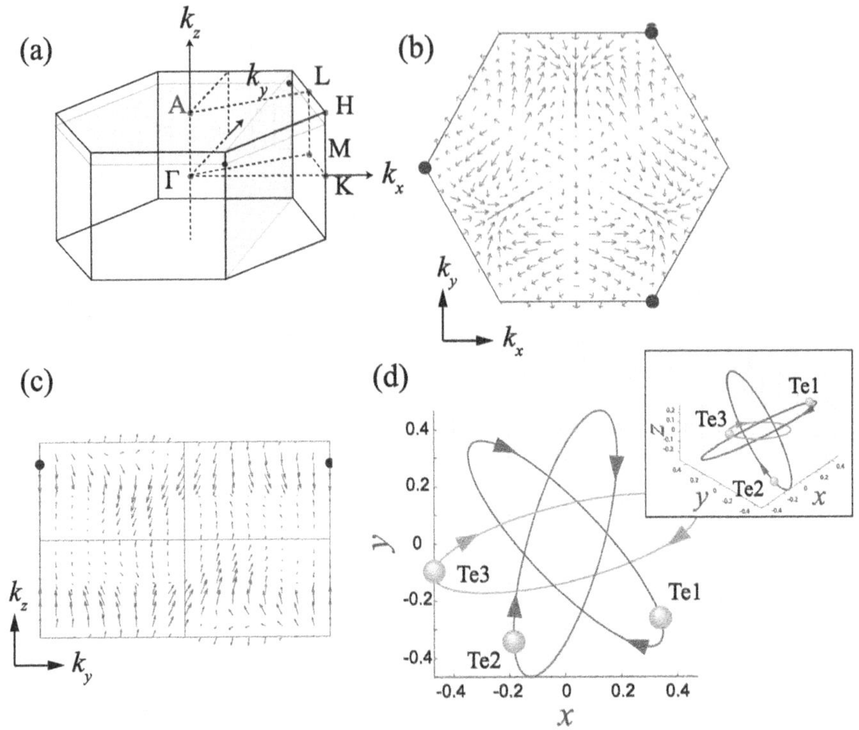


**Fig. 3.6** **b**, **c** Distribution of the phonon angular momentum $\mathbf{l}_\sigma(\mathbf{k})$ of the fourth lowest band. **b** and **c** show the results on the plane $\frac{k_z a}{2\pi} = 0.2927$ and on the plane $\frac{k_x a}{2\pi} = \frac{1}{3}$, respectively. These planes correspond to the two cross sections in **a**. **d** Trajectories of the four atoms in the unit cell for the phonon of the fourth lowest band at $\frac{\mathbf{k}a}{2\pi} = (\frac{1}{3}, \frac{1}{\sqrt{3}}, 0.2927)$, which is indicated as the black dots in **a**, **b**, and **c**. **d** represents the normalized polarization vector $\varepsilon_\sigma(\mathbf{k})$ with $\varepsilon^\dagger\varepsilon = 1$, and its axis $(x, y, z)$ is shown in a dimensionless unit. Reprinted figure with permission from [11] Masato Hamada, Emi Minamitani, Motoaki Hirayama, and Shuichi Murakami, "Phonon Angular Momentum Induced by the Temperature Gradient", Phys. Rev. Lett. 121. 175301, 2018, Copyright 2018 by the American Physical Society

as $\alpha_{zz} \sim -10^{-7} \times [\tau_\parallel/(1s)]$ $\text{Jsm}^{-2}\text{K}^{-1}$ and $\alpha_{xx} \sim 10^{-7} \times [\tau_\perp/(1s)]$ $\text{Jsm}^{-2}\text{K}^{-1}$ at $T = 300$ K, and that for Se is estimated as $\alpha_{zz} \sim -10^{-6} \times [\tau_\parallel/(1s)]$ $\text{Jsm}^{-2}\text{K}^{-1}$ and $\alpha_{xx} \sim -10^{-7} \times [\tau_\perp/(1s)]$ $\text{Jsm}^{-2}\text{K}^{-1}$ at $T = 300$ K. We show the numerical results for Te and Se by using the first-principle calculation in Table 3.2.

**Table 3.2** Numerical results of induced phonon angular momentum and a rigid-body rotation of Te and Se. Modified table with permission from [11] Masato Hamada, Emi Minamitani, Motoaki Hirayama, and Shuichi Murakami, "Phonon Angular Momentum Induced by the Temperature Gradient", Phys. Rev. Lett. 121. 175301, 2018, Copyright 2018 by the American Physical Society

| | Te | Se |
|---|---|---|
| $a$ [Å] | 4.456 | 4.371 |
| $c$ [Å] | 5.921 | 4.954 |
| $\rho$ [kg/m$^3$] | $6.245 \times 10^3$ | $4.800 \times 10^3$ |
| $\tau_{\parallel}$ [ps] | $\sim$10 | $\sim$10 |
| $\tau_{\perp}$ [ps] | $\sim$1 | $\sim$10 |
| $\alpha_{xx}$ [Jsm$^{-2}$K$^{-1}$] | $-3.0 \times 10^{-7} \times \tau_{\perp}$ | $-7.1 \times 10^{-7} \times \tau_{\perp}$ |
| $\alpha_{zz}$ [Jsm$^{-2}$K$^{-1}$] | $4.8 \times 10^{-7} \times \tau_{\parallel}$ | $-9.7 \times 10^{-6} \times \tau_{\parallel}$ |
| $\Omega_{\text{rigid-body}}$ perpendicular to c-axis [s$^{-1}$] | $2.9 \times 10^{-22} \times \Delta T/L^3$ | $8.9 \times 10^{-21} \times \Delta T/L^3$ |
| $\Omega_{\text{rigid-body}}$ along c-axis [s$^{-1}$] | $-4.6 \times 10^{-21} \times \Delta T/L^3$ | $1.2 \times 10^{-19} \times \Delta T/L^3$ |

## 3.4 Conversion of Phonon Angular Momentum Generated by Temperature Gradient

We have found that the phonon angular momentum is generated by temperature gradient. Nevertheless, the phonon angular momentum represents microscopic local rotations of the nuclei, and we cannot measure the phonon angular momentum directly. We consider conversion of the phonon angular momentum into other degree of freedom and propose experiments to measure this. In this section, we propose the conversion of the phonon angular momentum to a rigid-body rotation of the crystal and to a magnetization.

### *3.4.1 Conversion to a Rigid-Body Rotation of Crystal*

To measure the phonon angular momentum generated by temperature gradient, we consider the phonon version of the Einstein-de Haas effect. Suppose the crystal can rotate freely, and the crystal has no angular momentum. By conservation of angular momentum, when a heat current generates a phonon angular momentum $J^{\text{ph}}$, a rigid-body rotation of the crystal also acquires an angular momentum which compensates the phonon angular momentum, $J^{\text{rigid-body}} = -J^{\text{ph}}$. This conservation holds when we take an average over a long period much longer than a typical time scale of the phonon motions. For example, in polar crystals such as wurtzite GaN, when the heat current flows along the $y$ direction, the phonon angular momentum along the $x$ direction is generated, and it is converted to a rigid-body rotation, as shown in

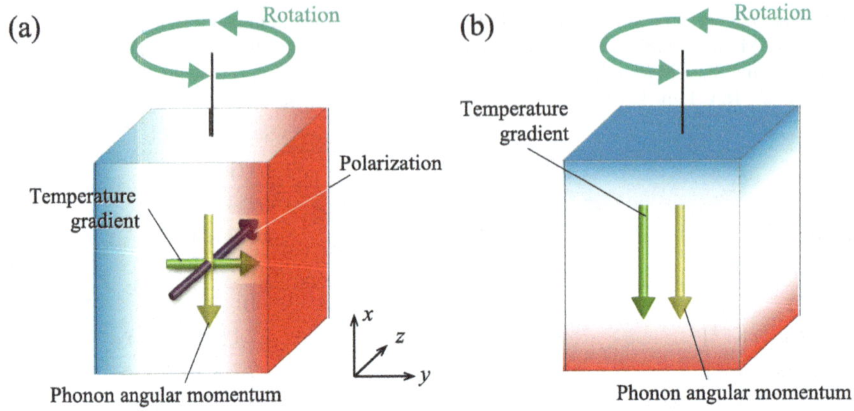


**Fig. 3.7** Schematic diagram of the generated rigid-body rotation due to the heat current. **a** and **b** shows two typical cases: **a** for a polar crystal such as wurtzite GaN and **b** for a chiral crystal such as Te and Se. Reprinted figure with permission from [11] Masato Hamada, Emi Minamitani, Motoaki Hirayama, and Shuichi Murakami, "Phonon Angular Momentum Induced by the Temperature Gradient", Phys. Rev. Lett. 121. 175301, 2018, Copyright 2018 by the American Physical Society

Fig. 3.7a. Similarly, in tellurium, it is schematically shown in Fig. 3.7b, and the rotation direction will be opposite for right-handed and left-handed crystals.

Next, we estimate the angular velocity $\Omega$ of the rigid-body rotation in GaN as an example. We set the sample size to be $L \times L \times L$ and the phonon relaxation time to be $\tau \sim 10$ ps [20]. The temperature difference over the sample size $L$ is denoted by $\Delta T$. The angular momentum of the rigid-body rotation is represented as $J^{\text{rigid-body}} L^3 = I\Omega$, where $I = ML^2/6$ is the inertial moment of the sample with the total mass $M$. We estimate the angular velocity of the rigid-body rotation as

$$\Omega = -J_x^{\text{ph}} L^3 / I \sim \frac{\Delta T/(1\text{K})}{(L/(1\text{m}))^3} \times 10^{-21}\ \text{s}^{-1}. \tag{3.22}$$

Then, by setting the temperature difference to be $\Delta T = 10$ K, an angular velocity of the rigid-body rotation $\Omega$ is estimated as $10^{-8}$ s$^{-1}$ when we set the sample size $L = 100$ μm and $10^{-2}$ s$^{-1}$ when we set the sample size $L = 1$ μm. They are sufficiently large for experimental measurement.

We also estimate the angular velocity $\Omega$ of the rigid-body rotation in Te and Se. We assume the phonon relaxation times along and perpendicular to c-axis of Te are $\tau_{\parallel} \sim 10$ ps and $\tau_{\perp} \sim 1$ ps, and those of Se are $\tau_{\parallel} \sim 10$ ps and $\tau_{\perp} \sim 10$ ps, respectively [22–24]. From Eq. (3.11), the nonzero components of response function are $\alpha_{xx} = \alpha_{yy}$ and $\alpha_{zz}$. Therefore, when we apply the temperature gradient along the $z$ axis to the sample, the phonon angular momentum along the $z$ axis is induced, and when we apply the temperature gradient along the $y$ axis to the sample, the phonon angular momentum is along the $y$ axis. For wurtzite GaN under the same condition, we also estimate the angular velocities of the rigid-body rotation of Te as

$$\Omega_{\perp} \sim \frac{\Delta T/(1\mathrm{K})}{(L/(1\mathrm{m}))^3} \times 10^{-22}\ \mathrm{s}^{-1}, \tag{3.23}$$

$$\Omega_{\parallel} \sim -\frac{\Delta T/(1\mathrm{K})}{(L/(1\mathrm{m}))^3} \times 10^{-21}\ \mathrm{s}^{-1}, \tag{3.24}$$

and these of Se as

$$\Omega_{\perp} \sim \frac{\Delta T/(1\mathrm{K})}{(L/(1\mathrm{m}))^3} \times 10^{-21}\ \mathrm{s}^{-1}, \tag{3.25}$$

$$\Omega_{\parallel} \sim \frac{\Delta T/(1\mathrm{K})}{(L/(1\mathrm{m}))^3} \times 10^{-19}\ \mathrm{s}^{-1}. \tag{3.26}$$

Then, for the temperature difference $\Delta T = 10$ K, the angular velocities of the rigid-body rotation $\Omega_{\perp}$ and $\Omega_{\parallel}$ of Te are estimated as $10^{-9}\ \mathrm{s}^{-1}$ and $-10^{-8}\ \mathrm{s}^{-1}$ when we set the sample size $L = 100\,\mu\mathrm{m}$, and those of Se are estimated as $10^{-8}\ \mathrm{s}^{-1}$ and $-10^{-6}\ \mathrm{s}^{-1}$ when we set the sample size $L = 100\ \mu\mathrm{m}$. In this setup, the induced phonon angular momentum and a rigid-body rotation are expected to be experimentally observable, and the induced phonon angular momentum in Se is larger than that of Te. We discuss reason why the sum of the angular momentum is conserved in the present case. Here, let $\mathbf{R}_j$ denote the equilibrium position of the $j$-th atom, and and $\mathbf{u}_j$ denote its deviation from the equilibrium position. We define the mass of the $j$-th atom $m_j$. Then the conservation of angular momentum is written as $\sum_j m_j(\mathbf{R}_j + \mathbf{u}_j) \times (\dot{\mathbf{R}}_j + \dot{\mathbf{u}}_j) = 0$. We average this quantity over a long period, which is much longer than the typical timescale of the phonon motions. Then, the typical time scale of $\mathbf{R}$ is $10^2$–$10^8$ s as we estimate for GaN, Te and Se, and it is much longer than the time scale of the phonon motion. Therefore, the time-averages of the cross terms $\sum_j m_j \mathbf{R}_j \times \dot{\mathbf{u}}_j$, and $\sum_j m_j \mathbf{u}_j \times \dot{\mathbf{R}}_j = 0$ vanish and are neglected. Then after time-averaging, the sum of $J^{\mathrm{ph}} = \sum_j m_j \mathbf{u}_j \times \dot{\mathbf{u}}_j$, and $J^{\mathrm{rigid-body}} = \sum_j m_j \mathbf{R}_j \times \dot{\mathbf{R}}_j$ is zero. Therefore, the phonon angular momentum is converted to a rigid-body rotation.

### 3.4.2 Conversion to a Magnetization

We also propose the conversion from the phonon angular momentum to a magnetization. Polar crystals like wurtzite GaN have nuclei with effective charges. When a temperature gradient is applied to polar crystals, the phonon angular momentum is generated and the nuclei with effective charges rotate around their equilibrium positions microscopically. Therefore, it induces magnetization in itself. The magnetic moment $\mathbf{m}$ and the angular momentum are represented by $\mathbf{m} = \gamma \mathbf{j}$ with the gyromagnetic ratio $\gamma$. In wurtzite GaN, the Born effective charge tensor $eZ^*_{\alpha\beta}$ is $eZ^*_{xx} = eZ^*_{yy} = 2.58e, eZ^*_{zz} = 2.71e$ from our first-principle calculation, where $e$ is an electron charge. The ratio tensors of the Ga and N atoms are given by $\gamma^{\mathrm{Ga}}_{\alpha\beta} = geZ^*_{\alpha\beta}/2m_{\mathrm{Ga}}$ and $\gamma^{\mathrm{N}}_{\alpha\beta} = -geZ^*_{\alpha\beta}/2m_{\mathrm{N}}$ with g-factor of GaN $g_{\parallel} = 1.951, g_{\perp} = 1.9483$ [25], where $m_{\mathrm{Ga}}$ and $m_{\mathrm{N}}$ are the mass of the Ga atom and that of the N atom, respectively. We estimate the order of magnitude of the magnetization as

$$M_x \sim \frac{\Delta T/(1\mathrm{K})}{L/(1\mathrm{m})} \times 10^{-11}\ \mathrm{Am}^{-1}. \tag{3.27}$$

Therefore the magnetization along the $x$ direction $M_x$ of GaN induced by temperature gradient is estimated as $10^{-6}$ $\mathrm{Am}^{-1}$ when we set the sample size $L = 100$ μm and the temperature difference $\Delta T = 10$ K and $10^{-4}$ $\mathrm{Am}^{-1}$ when we set $L = 1$ μm and $\Delta T = 10$ K. Although the order of magnitude of this magnetization is very small, it is expected to be observable experimentally.

## 3.5 Phonon Thermal Edelstein Effect in the High- and Low-Temperature Limits

In this section, we discuss the temperature dependence of the phonon angular momentum generated by the temperature gradient [26]. We focus on the high-temperature limit and the low-temperature limit.

First, we discuss the high-temperature limit. At high temperature $\hbar\omega_\sigma(\mathbf{k}) \ll k_{\mathrm{B}}T$, we can expand the Bose distribution function in Taylor series as

$$\begin{aligned} f_0(x) &= \frac{1}{e^x - 1} \\ &= \frac{1}{x}\left[1 - \left(\frac{1}{2}x + \frac{1}{6}x^2 + \cdots\right) + \left(\frac{1}{2}x + \frac{1}{6}x^2 + \cdots\right)^2 - \left(\frac{1}{2}x + \frac{1}{6}x^2 + \cdots\right)^3 + \cdots\right] \\ &= \frac{1}{x} - \frac{1}{2} + \frac{1}{12}x + O(x^2) \end{aligned} \tag{3.28}$$

with $x = \frac{\hbar\omega}{k_{\mathrm{B}}T}$. Then, the temperature derivative of the Bose distribution function is represented as

$$\frac{\partial f_0(x)}{\partial T} = -\frac{\hbar\omega}{k_{\mathrm{B}}T^2}\left(-\frac{1}{x^2} + \frac{1}{12} + O(x)\right) \simeq \frac{k_{\mathrm{B}}}{\hbar\omega} - \frac{\hbar\omega}{12k_{\mathrm{B}}T^2}. \tag{3.29}$$

The response tensor $\alpha$ in the high-temperature limit is represented as

$$\alpha_{ij} = -\frac{\tau}{V}\sum_{\mathbf{k},\sigma} l_{\sigma,i}(\mathbf{k}) v_{\sigma,j}(\mathbf{k}) \frac{k_{\mathrm{B}}}{\hbar\omega_\sigma(\mathbf{k})}. \tag{3.30}$$

This shows that the response tensor becomes constant in the high-temperature limit.

Next, we discuss the low-temperature limit. Because populations in the phonon modes except for the acoustic modes are negligibly small at low temperature, we consider only the acoustic modes with a long wavelength. In the long wavelength limit, the frequencies of acoustic phonon modes are represented as

$$\omega_\sigma(\mathbf{k}) = v_\sigma \sqrt{k_x^2 + k_y^2 + k_z^2} = v_\sigma k, \tag{3.31}$$

where $\sigma = 1, 2, 3$ represents a band index for the acoustic phonon modes. For simplicity, we assume that the group velocity is isotropic $v_{\sigma,x} = v_{\sigma,y} = v_{\sigma,z} = v_\sigma$ and we change the wave vector $\mathbf{k}$ from the Cartesian coordinate $(k_x, k_y, k_z)$ to the polar coordinate $(k, \theta, \phi)$. We can expand the phonon angular momentum of each mode in the Taylor series as

$$l_{\sigma,i}(\mathbf{k}) = \beta_{\sigma,ij} k_j + O(k^3). \tag{3.32}$$

Since the phonon angular momentum of each mode is an odd function of $\mathbf{k}$ in the system without inversion symmetry, it can be expanded into odd powers in the wave vector $\mathbf{k}$. The coefficient tensor $\beta_{ij}$ is an axial tensor because the phonon angular momentum of each mode is an axial vector and the wave vector is a polar vector. In the system of an infinite size, the wave vectors are continuous, so that one can replace summation by integration,

$$\frac{1}{V}\sum_{\mathbf{k}} \to \int_{\mathrm{B.Z.}} \frac{d^3k}{(2\pi)^3} = \frac{1}{(2\pi)^3}\int_0^{2\pi} d\phi \int_0^{\pi} d\theta \int_0^{k^c} dk k^2 \sin\theta, \tag{3.33}$$

where $k^c$ is a cutoff wavenumber introduced for the short wavelength. Then, the response tensor $\alpha_{xy}$ can be rewritten as

$$\begin{aligned}
\alpha_{xy} &= -\frac{\iota}{V}\sum_{\sigma=1,2,3}\sum_{\mathbf{k}} l_{\sigma,x}(\mathbf{k}) v_{\sigma,y} \frac{\partial f(\omega_\sigma(k))}{\partial T} \\
&= \sum_{\sigma=1,2,3} \frac{-\tau v_\sigma}{(2\pi)^3} \int_0^{2\pi} d\phi \int_0^{\pi} d\theta \int_0^{k^c} dk k^3 \sin^2\theta \sin\phi \\
&\qquad \times (\beta_{\sigma,xx}\sin\theta\cos\phi + \beta_{\sigma,xy}\sin\theta\sin\phi + \beta_{\sigma,xz}\cos\theta) \frac{\partial f(\omega_\sigma(k))}{\partial T} \\
&= \sum_{\sigma=1,2,3} \frac{-\pi\tau v_\sigma \beta_{\sigma,xy}}{(2\pi)^3} \int_0^{\pi} d\theta \int_0^{k^c} dk k^3 \sin^3\theta \frac{\partial f(\omega_\sigma(k))}{\partial T} \\
&= \sum_{\sigma=1,2,3} \frac{-\tau v_\sigma \beta_{\sigma,xy}}{6\pi^2} \int_0^{k^c} dk k^3 \frac{\partial f(\omega_\sigma(k))}{\partial T}.
\end{aligned} \tag{3.34}$$

Here, we use the following relation

$$\frac{\partial f(\omega_\sigma(k))}{\partial T} = -\frac{\omega_\sigma}{T}\frac{\partial f(\omega_\sigma)}{\partial \omega_\sigma}, \tag{3.35}$$

and we introduce the cutoff frequency $\omega_\sigma^c = v_\sigma k^c$. Then, the response tensor is represented as

$$\begin{aligned}
\alpha_{xy} &= \sum_{\sigma=1,2,3} \frac{\tau v_\sigma \beta_{\sigma,xy}}{6\pi^2 T} \int_0^{k^c} dk k^3 \omega_\sigma \frac{\partial f(\omega_\sigma)}{\partial \omega_\sigma} \\
&= \sum_{\sigma=1,2,3} \frac{\tau \beta_{\sigma,xy}}{6\pi^2 v_\sigma^3 T} \int_0^{\omega_\sigma^c} d\omega_\sigma \omega_\sigma^4 \frac{\partial f(\omega_\sigma)}{\partial \omega_\sigma} \\
&= \sum_{\sigma=1,2,3} \frac{\tau \beta_{\sigma,xy}}{6\pi^2 v_\sigma^3 T} \left[ \omega_\sigma^{c4} f(\omega_\sigma^c) - 4 \int_0^{\omega_\sigma^c} d\omega_\sigma \omega_\sigma^3 f(\omega_\sigma) \right].
\end{aligned} \tag{3.36}$$

Because $x_\sigma = \hbar\omega_\sigma/(k_B T) \to \infty$ in the low-temperature limit, the second term can be calculated as

$$\begin{aligned}
\int_0^{\omega_\sigma^c} d\omega_\sigma \omega_\sigma^3 f(\omega_\sigma) &\cong \frac{(k_B T)^4}{\hbar^4} \int_0^\infty dx_\sigma \frac{x_\sigma^3}{e^{x_\sigma} - 1} \\
&= \frac{(k_B T)^4}{\hbar^4} \int_0^\infty dx_\sigma \frac{x_\sigma^3 e^{-x_\sigma}}{1 - e^{-x_\sigma}} \\
&= \frac{(k_B T)^4}{\hbar^4} \int_0^\infty dx_\sigma \sum_{s=1}^\infty x_\sigma^3 e^{-s x_\sigma} \\
&= \frac{(k_B T)^4}{\hbar^4} 6\zeta(4) \\
&= \frac{\pi^4 k_B^4}{15\hbar^4} T^4.
\end{aligned} \tag{3.37}$$

The first term of Eq. (3.36) vanishes in the low-temperature limit. Therefore, the response tensor $\alpha_{xy}$ in the low-temperature limit is represented as

$$\alpha_{xy} = -\sum_{\sigma=1,2,3} \frac{2\pi^2 \tau \beta_{xy,\sigma} k_B^4}{45 v_\sigma^3 \hbar^4} T^3. \tag{3.38}$$

The temperature dependence of other components can be expressed in the same way, and they are proportional to $T^3$ in low temperature limit.

We discuss dependence of the phonon thermal Edelstein effect as the mass $m$ of the nuclei. Suppose we change the mass of the nuclei, without changing the lattice structure and the elastic constants between the nuclei. In the continuum limit, the group velocity is proportional to $m^{-\frac{1}{2}}$, and the magnitude of the phonon angular momentum does not depend on the mass. Therefore, the coefficient of $\alpha/T^3$ is proportional to $m^{\frac{3}{2}}$. However, at finite temperature, the energy of phonon depend on mass, and we cannot compare simplicity the phonon angular momentum for mass.

## 3.6 Summary

In this chapter, we have theoretically predicted the phonon angular momentum generated by the heat current, and we have estimated the phonon angular momentum for wurtzite GaN, Te, and Se by using the first-principle calculation. This mechanism is analogous to the Edelstein effect in electronic systems, and we call it the phonon thermal Edelstein effect. We proposed the two methods to measure the phonon angular momentum generated by the heat current. When the crystals can rotate freely and the heat current is applied, the phonon angular momentum generated by heat current is converted to a rigid-body rotation of the crystals due to the conservation of angular momentum. This rigid-body rotation is sufficiently large for experimental measurement when the size of sample is micro order. On the other hand, because of the nuclei having positive charges, the phonon angular momentum generated by heat current induces magnetization. The response tensor for the phonon thermal Edelstein effect is proportional to $T^3$ in low-temperature limit, and this converges to a constant in high-temperature limit.

Moreover, in metals, the phonon angular momentum will be partially converted to electronic spin angular momentum via the spin-rotation coupling, which is similar to the spin-current generation proposed for the surface acoustic waves in solids [27], and for the twiston modes in carbon nanotubes [28]. These experimental proposals are expected to unveil properties of the phonon angular momentum.

## References


1. Zhang L, Niu Q (2014) Angular momentum of phonons and the einstein-de haas effect. Phys Rev Lett 112:085503
2. Srivastava G (1990) The physics of phonons. Taylor & Francis
3. Ibach H, Lüth H (2003) Solid-state physics: an introduction to principles of materials science ; with 100 problems. Advanced texts in physics. Springer
4. Edelstein VM (1990) Spin polarization of conduction electrons induced by electric current in two-dimensional asymmetric electron systems. Solid State Commun 73:233–235
5. Ivchenko EL, Pikus GE (1978) New photogalvanic effect in gyrotropic crystals 27:604
6. Levitov LS, Nazarov YV, Eliashberg GM (1985) Magnetoelectric effects in conductors with mirror isomer symmetry 61:133
7. Aronov AG, Lyanda-Geller YB (1989) Nucleasr electric resonance and orientation of carier spins by an electric field 50:431
8. Inoue J-I, Bauer GEW, Molenkamp LW (2003) Diffuse transport and spin accumulation in a rashba two-dimensional electron gas. Phys Rev B 67:033104
9. Kato YK, Myers RC, Gossard AC, Awschalom DD (2004) Current-induced spin polarization in strained semiconductors. Phys Rev Lett 93:176601
10. Yoda T, Yokoyama T, Murakami S (2015) Current-induced orbital and spin magnetizations in crystals with helical structure. Sci Rep 5:12024
11. Hamada M, Minamitani E, Hirayama M, Murakami S (2018) Phonon angular momentum induced by the temperature gradient. Phys Rev Lett 121:175301. https://link.aps.org/doi/10.1103/PhysRevLett.121.175301
12. Birss RR (1962) In: Wohlfarth EP (ed) Symmetry and magnetism, vol 3. Series of monographs on selected topics in solid state physics. Elsevier North-Holland

13. Camacho D, Niquet Y (2010) Application of keating's valence force field model to non-ideal wurtzite materials. Phys E: Low-Dimens Syst Nanostructures 42:1361–1364
14. Baroni S, de Gironcoli S, Dal Corso A, Giannozzi P (2001) Phonons and related crystal properties from density-functional perturbation theory. Rev Mod Phys 73:515–562
15. Giannozzi P et al (2009) Quantum espresso: a modular and open-source software project for quantum simulations of materials. J Phys Condens Matter 21:395502
16. Garrity KF, Bennett JW, Rabe KM, Vanderbilt D (2014) Pseudopotentials for high-throughput dft calculations. Comput Mater Sci 81:446–452
17. Monkhorst HJ, Pack JD (1976) Special points for brillouin-zone integrations. Phys Rev B 13:5188–5192
18. Ruf T et al (2001) Phonon dispersion curves in wurtzite-structure gan determined by inelastic x-ray scattering. Phys Rev Lett 86:906–909
19. Bungaro C, Rapcewicz K, Bernholc J (2000) Ab initio. Phys Rev B 61:6720–6725
20. Wu X et al (2016) Thermal conductivity of wurtzite zinc-oxide from first-principle lattice dynamics - a comparative study with gallium nitride. Sci Rep 6:22504
21. Hirayama M, Okugawa R, Ishibashi S, Murakami S, Miyake T (2015) Weyl node and spin texture in trigonal tellurium and selenium. Phys Rev Lett 114:206401
22. Cooper CW (ed) (1969) The physics of selenium and tellurium. Pergamon
23. Gerlach E, Grosse P (eds) (1979) The physics of selenium and tellurium, vol 13. Springer series in solid-state sciences. Springer, Berlin Heidelberg
24. Peng H, Kioussis N, Stewart DA (2015) Anisotropic lattice thermal conductivity in chiral tellurium from first principles. Appl Phys Lett 107:251904
25. Carlos WE, Freitas JA, Khan MA, Olson DT, Kuznia JN (1993) Electron-spin-resonance studies of donors in wurtzite gan. Phys Rev B 48:17878–17884
26. Hamada M, Murakami S (2020) Phonon rotoelectric effect. Phys Rev B 101:144306. https://link.aps.org/doi/10.1103/PhysRevB.101.144306
27. Matsuo M, Ieda J, Harii K, Saitoh E, Maekawa S (2013) Mechanical generation of spin current by spin-rotation coupling. Phys Rev B 87:180402
28. Hamada M, Yokoyama T, Murakami S (2015) Spin current generation and magnetic response in carbon nanotubes by the twisting phonon mode. Phys Rev B 92:060409

# Chapter 4
# Magnetoelectric Effect for Phonons

## 4.1 Phonon Angular Momentum with Neither Time-Reversal nor Inversion Symmetries

The phonon angular momentum for each mode $\mathbf{l}_\sigma$ is an odd (even) function of phonon wave vector $\mathbf{k}$ when the system has time-reversal (inversion) symmetry. Thus, to make phonon angular momentum nonzero $\mathbf{l}_\sigma(\mathbf{k}) \neq 0$, at least one of these symmetries must be broken. In this chapter, we consider systems with neither of these symmetries, for example, magnetic crystals.

First, we consider the phonon thermal Edelstein effect extended to the systems neither time-reversal nor inversion symmetries. When the temperature gradient is applied to a system with neither of these symmetries, the phonon angular momentum consists of two terms: an equilibrium term and a term proportional to the temperature gradient. Because the phonon angular momentum of each mode is no longer an even or odd function of the wave vector $\mathbf{k}$ in the system neither time-reversal nor inversion symmetries, the phonon angular momentum in equilibrium does not cancel between $\mathbf{k}$ and $-\mathbf{k}$. In addition, the phonon thermal Edelstein effect discussed in the previous chapter is also allowed. Note that the response tensor for the phonon thermal Edelstein effect in magnetic crystals is determined not by their point groups but by their magnetic point groups.

On the other hand, when both symmetries are broken and their product is conserved, the phonon angular momentum for each mode $\mathbf{l}_\sigma(\mathbf{k})$ becomes zero at any wave vector. Therefore, the total phonon angular momentum becomes zero. In this case, the phonon thermal Edelstein effect does not occur because the phonon angular momentum for each mode $\mathbf{l}_\sigma(\mathbf{k})$ is zero. Here, we introduce another mechanism of generation of the phonon angular momentum. A similar mechanism is known in multiferroics, called magnetoelectric effect [1]. In fact, the symmetry of this system is the same as that of multiferroic materials where the magnetoelectric effect occurs. We show that the phonon angular momentum is generated by an electric field, in analogy to the magnetoelectric effect [2]. In order to break the time-reversal symmetry in the phonon system, we introduce the spin-phonon interaction [3–7], which



M. Hamada, *Theory of Generation and Conversion of Phonon Angular Momentum*,
Springer Theses, https://doi.org/10.1007/978-981-33-4690-1_4

couples localized spins and the phonon angular momentum, and this is represented as

$$H_{\mathrm{SPI}} = -g \sum_{l\kappa} \mathbf{S}_\kappa \cdot (\mathbf{u}_{l\kappa} \times m_\kappa \dot{\mathbf{u}}_{l\kappa}), \tag{4.1}$$

where $g$ is a coupling constant, and $\mathbf{S}_\kappa$ is the magnetization of localized spins of the $\kappa$th atom in the unit cell. This terms means that the coupling of the charged ions to the magnetic field generated by the localized spins. Since this interaction Eq. (4.1) works like a Lorentz force to the charged ions, the motion of the ion is deflected and the phonon angular momentum is generated. By an analogy with magnetoelectric effect, when an electric field is applied, the phonon angular momentum is generated, and this is represented as

$$J^{\mathrm{ph}}_\alpha = \alpha_{\alpha\beta} E_\beta, \tag{4.2}$$

where $\mathbf{E}$ is an electric field and $\alpha$ is a response tensor. We call this the phonon rotoelectric effect since the microscopic local rotations of atoms are induced by the electric field [2]. The response tensor $\alpha$ is an axial tensor, and this response tensor has nonzero elements for 18 magnetic point groups with the product of time-reversal and inversion symmetries and 40 magnetic point groups without it. A full list of the form of the axial tensor $\alpha_{\alpha\beta}$ is available from Ref. [8]. We note that the phonon rotoelectric effect automatically induces a magnetization by phonons in ionic crystals. Moreover, in the phonon thermal Edelstein effect, the difference from the Bose distribution function is important, and in the phonon rotoelectric effect, the change of the phonon angular momentum for each mode is important.

## 4.2 Formalism of Phonons with the Spin-Phonon Interaction

In this section, we introduce a Lagrangian for phonons with spin-phonon interaction and derive the equation of the motion for phonons. The Lagrangian for the phonon system with the spin-phonon interaction is represented as

$$L = \sum_{l\kappa} \left[ \frac{m_\kappa}{2} \dot{\mathbf{u}}^2_{l\kappa} - \frac{1}{2} \sum_{l'\kappa'} \Phi_{\alpha\beta}(l\kappa, l'\kappa') u_{l\kappa,\alpha} u_{l'\kappa',\beta} - g\mathbf{S}_\kappa \cdot [\mathbf{u}_{l\kappa} \times m_\kappa \dot{\mathbf{u}}_{l\kappa}] \right], \tag{4.3}$$

where $m_\kappa$ is the mass of the $\kappa$th atom in the unit cell, $\Phi$ is a force constant matrix, and $\mathbf{u}_{l\kappa}$ is displacement vector of the $\kappa$th atom in the $l$th unit cell. The first term is a kinetic term, the second term is a potential term, and the third term is a spin-phonon interaction term. To make it compact, we rewrite the displacement vector within the unit cell into a $3N$ dimensional vector represented as

$$\mathbf{u}_{l\kappa} \rightarrow u_l = (\sqrt{m_1}u_{l1,x}, \sqrt{m_1}u_{l1,y}, \sqrt{m_1}u_{l1,z}, \sqrt{m_2}u_{l2,x}, \ldots, \sqrt{m_N}u_{lN,z})^T, \quad (4.4)$$

where $N$ is the number of atoms per unit cell. The first term and the second term of Eq. (4.3) are rewritten as

$$\sum_{l\kappa} \frac{m_\kappa}{2}\dot{\mathbf{u}}_{l\kappa}^2 = \sum_l \frac{1}{2}\dot{u}_l^T \dot{u}_l, \quad (4.5)$$

$$\sum_{l\kappa,l'\kappa'} \frac{1}{2}\Phi_{\alpha\beta}(l\kappa, l'\kappa')u_{l\kappa,\alpha}u_{l'\kappa',\beta} = \sum_{ll'} \frac{1}{2}u_l^T \Phi_{ll'} u_{l'}. \quad (4.6)$$

The third term of Eq. (4.3) is also rewritten as a matrix form. First, the phonon angular momentum of the $\kappa$th atom is rewritten as

$$\begin{aligned}(\mathbf{u}_{l\kappa} \times m_\kappa \dot{\mathbf{u}}_{l\kappa})_x &= (\sqrt{m_\kappa}u_{l\kappa,y})(\sqrt{m_\kappa}\dot{u}_{l\kappa,z}) - (\sqrt{m_\kappa}u_{l\kappa,z})(\sqrt{m_\kappa}\dot{u}_{l\kappa,y}) \\ &= \left(\sqrt{m_\kappa}u_{l\kappa,x}\ \sqrt{m_\kappa}u_{l\kappa,y}\ \sqrt{m_\kappa}u_{l\kappa,z}\right)\begin{pmatrix}0 & 0 & 0\\ 0 & 0 & 1\\ 0 & -1 & 0\end{pmatrix}\begin{pmatrix}\sqrt{m_\kappa}\dot{u}_{l\kappa,x}\\ \sqrt{m_\kappa}\dot{u}_{l\kappa,y}\\ \sqrt{m_\kappa}\dot{u}_{l\kappa,z}\end{pmatrix} \\ &= u_{l\kappa}^T M_x \dot{u}_{l\kappa}. \end{aligned} \quad (4.7)$$

Similarly, the $y$ and $z$ components of the phonon angular momentum are also written as

$$(\mathbf{u}_{l\kappa} \times m_\kappa \dot{\mathbf{u}}_{l\kappa})_y = u_{l\kappa}^T M_y \dot{u}_{l\kappa}, \quad (4.8)$$

$$(\mathbf{u}_{l\kappa} \times m_\kappa \dot{\mathbf{u}}_{l\kappa})_z = u_{l\kappa}^T M_z \dot{u}_{l\kappa}, \quad (4.9)$$

with $(M_\alpha)_{\beta\gamma} = \varepsilon_{\alpha\beta\gamma}$. Thus, the third term of Eq. (4.3) is rewritten as

$$g\sum_{l\kappa} \mathbf{S}_\kappa \cdot [\mathbf{u}_{l\kappa} \times \dot{\mathbf{u}}_{l\kappa}] = \sum_l g u_l^T \begin{pmatrix} S_1 & 0 & 0 & \ldots & 0\\ 0 & S_2 & 0 & \ldots & 0\\ \vdots & \vdots & \vdots & \ddots & \vdots\\ 0 & 0 & 0 & \ldots & S_N\end{pmatrix}_\alpha \otimes M_\alpha \dot{u}_l \quad (4.10)$$

$$= \sum_{l,\alpha} u_l^T \beta_\alpha \dot{u}_l. \quad (4.11)$$

Therefore, the Lagrangian is rewritten as

$$L = \sum_l \left[\frac{1}{2}\dot{u}_l^T \dot{u}_l - \frac{1}{2}\sum_{l'} u_l^T \Phi_{ll'} u_{l'} - \sum_\alpha u_l^T \beta_\alpha \dot{u}_l\right]. \quad (4.12)$$

Note that the matrix $\beta_\alpha$ is an antisymmetric tensor $\beta_\alpha = -\beta_\alpha^T$. The canonical momentum $p_l = \partial L/\partial \dot{u}_l$ is represented as

$$p_l = \dot{u}_l + \sum_{\alpha} \beta_\alpha u_l, \tag{4.13}$$

and the Hamiltonian $H = \sum_l p_l^T \dot{u}_l - L$ is represented as

$$H = \sum_{l,\alpha} \left[ \frac{1}{2}(p_l - \beta_\alpha u_l)^T (p_l - \beta_\alpha u_l) + \frac{1}{2} \sum_{l'} u_l^T \Phi_{ll'} u_l \right]. \tag{4.14}$$

To calculate the phonon angular momentum, we need to obtain the phonon dispersion and the eigenmodes of the phonons. Here we introduce the Schrödinger-like equation for phonons in the system without time-reversal symmetry [9]. The Hamiltonian of the phonons is rewritten as

$$H = \sum_{l,\alpha} \left[ \frac{1}{2} \dot{u}_l^T \dot{u}_l + \frac{1}{2} \sum_{l'} u_l^T \Phi_{ll'} u_{l'} \right] \tag{4.15}$$

with $\dot{u}_l = p_l - \sum_\alpha \beta_\alpha u_l$. Let us define a vector in an extended coordinate-velocity space $(y_{l1}, y_{l2}, \ldots, y_{2dN}) = (u_l^T, \dot{u}_l^T)$. The Hamiltonian is represented in the extended coordinate-velocity space as

$$H = \frac{1}{2} \sum_{l,l'} y_l^T Q_{ll'} y_{l'}, \tag{4.16}$$

$$Q_{ll'} = \begin{pmatrix} \Phi_{ll'} & 0 \\ 0 & I_{dN \times dN} \delta_{ll'} \end{pmatrix}, \tag{4.17}$$

with $d$ being the dimension of the system. Next, we introduce the Poisson bracket $\{A, B\} = \sum_i (\frac{\partial A}{\partial u_i} \frac{\partial B}{\partial p_i} - \frac{\partial A}{\partial p_i} \frac{\partial B}{\partial u_i})$, and calculate $\{y_i, y_j\}$ as follows

$$\begin{aligned} \{u_i, u_j\} &= \sum_k \left( \frac{\partial u_i}{\partial u_k} \frac{\partial u_j}{\partial p_k} - \frac{\partial u_i}{\partial p_k} \frac{\partial u_j}{\partial u_k} \right) \\ &= \sum_k (\delta_{ik} 0 - 0 \delta_{jk}) = 0, \end{aligned} \tag{4.18}$$

$$\begin{aligned} \{u_i, \dot{u}_j\} &= \sum_k \left( \frac{\partial u_i}{\partial u_k} \frac{\partial \dot{u}_j}{\partial p_k} - \frac{\partial u_i}{\partial p_k} \frac{\partial \dot{u}_j}{\partial u_k} \right) \\ &= \sum_k \left( \frac{\partial u_i}{\partial u_k} \frac{\partial (p_j - \beta_{j\mu} u_\mu)}{\partial p_k} - \frac{\partial u_i}{\partial p_k} \frac{\partial (p_j - \beta_{j\mu} u_\mu)}{\partial u_k} \right) \\ &= \delta_{ij}, \end{aligned} \tag{4.19}$$

$$\{\dot{u}_i, u_j\} = \sum_k \left( \frac{\partial \dot{u}_i}{\partial u_k} \frac{\partial u_j}{\partial p_k} - \frac{\partial \dot{u}_i}{\partial p_k} \frac{\partial u_j}{\partial u_k} \right)$$

$$= \sum_k \left( \frac{\partial(p_i - \beta_{i\mu}u_\mu)}{\partial u_k}\frac{\partial u_j}{\partial p_k} - \frac{\partial(p_i - \beta_{i\mu}u_\mu)}{\partial p_k}\frac{\partial u_j}{\partial u_k} \right)$$

$$= -\delta_{ij}, \tag{4.20}$$

$$\{\dot{u}_i, \dot{u}_j\} = \sum_k \left( \frac{\partial \dot{u}_i}{\partial u_k}\frac{\partial \dot{u}_j}{\partial p_k} - \frac{\partial \dot{u}_i}{\partial p_k}\frac{\partial \dot{u}_j}{\partial u_k} \right)$$

$$= \sum_k \left( \frac{\partial(p_i - \beta_{i\mu}u_\mu)}{\partial u_k}\frac{\partial(p_j - \beta_{j\mu}u_\mu)}{\partial p_k} - \frac{\partial(p_i - \beta_{i\mu}u_\mu)}{\partial p_k}\frac{\partial(p_j - \beta_{j\mu}u_\mu)}{\partial u_k} \right)$$

$$= \sum_k (-\beta_{ik}\delta_{jk} - \delta_{ik}(-\beta_{jk}))$$

$$= -2\beta_{ij}. \tag{4.21}$$

Thus, the Poisson bracket $\{y_{li}, y_{l'j}\}$ in the extended coordinate-velocity space is represented as

$$\{y_{li}, y_{l'j}\} = -\mathrm{i}R_{ij}\delta_{ll'}, \quad R = \begin{pmatrix} 0 & \mathrm{i}I \\ -\mathrm{i}I & -2\mathrm{i}\beta \end{pmatrix}. \tag{4.22}$$

The equation of motion $\dot{y}_{li} = \{y_{li}, H\}$ is represented as

$$\partial_t u_{li} = \{u_{li}, H\}$$

$$= \left\{ u_{li}, \frac{1}{2}\sum_l \dot{u}_{lk}\dot{u}_{lk} + \frac{1}{2}\sum_{l,l'} \Phi_{l\mu,l'\nu}u_{l\mu}u_{l'\nu} \right\}$$

$$= \dot{u}_{li}, \tag{4.23}$$

$$\partial_t \dot{u}_{li} = \{\dot{u}_{li}, H\}$$

$$= \left\{ \dot{u}_{li}, \frac{1}{2}\sum_l \dot{u}_{lk}\dot{u}_{lk} + \frac{1}{2}\sum_{l,l'} \Phi_{l\mu,l'\nu}u_{l\mu}u_{l'\nu} \right\}$$

$$= -2\beta_{i\mu}\dot{u}_\mu - \sum_{l'} \Phi_{li,l'\mu}u_{l'\mu}, \tag{4.24}$$

$$\partial_t \begin{pmatrix} u_l \\ \dot{u}_l \end{pmatrix} = \begin{pmatrix} O & I\delta_{ll'} \\ -\sum_{l'} \Phi_{ll'} & -2\beta\delta_{ll'} \end{pmatrix} \begin{pmatrix} u_{l'} \\ \dot{u}_{l'} \end{pmatrix}$$

$$= -\mathrm{i}R\sum_{l'} Q_{ll'} \begin{pmatrix} u_{l'} \\ \dot{u}_{l'} \end{pmatrix}, \tag{4.25}$$

and this is simplified as

$$\mathrm{i}R^{-1}\dot{y}_l = \sum_{l'} Q_{ll'}y_{l'}. \tag{4.26}$$

By assuming $u_l = \sum_{\mathbf{k}} \epsilon_{\mathbf{k}} e^{\mathrm{i}\mathbf{k}\cdot\mathbf{R}_l - \mathrm{i}\omega t}$, the above equation is changed into a generalized Hermitian eigenvalue problem

$$\sum_{l'} Q_{ll'} y_{l'} = \omega R^{-1} y_l \tag{4.27}$$

where both $R$ and $Q$ are Hermitian. The matrix $Q$ is semi-positive-definite as required by the structural stability of the system, and thus $Q^{1/2}$ is Hermitian. We introduce another vector $\psi_l$ in the extended coordinate-velocity space as

$$\psi_l = \sum_{l'} Q_{ll'}^{1/2} y_{l'} = \begin{pmatrix} \sum_{l'} \Phi_{ll'}^{1/2} u_{l'} \\ \dot{u}_l \end{pmatrix}, \tag{4.28}$$

and then we get a Schrödinger-like equation of phonons in the real space: $H\psi_l = \omega\psi_l$, where the Hamiltonian is

$$H = \sum_{l'} Q_{ll'}^{1/2} R Q_{ll'}^{1/2} = \begin{pmatrix} 0 & \mathrm{i}\sum_{l'} \Phi_{ll'}^{1/2} \\ -\mathrm{i}\sum_{l'} \Phi_{ll'}^{1/2} & -2\mathrm{i}\beta \end{pmatrix}. \tag{4.29}$$

In $k$-space, the Schrödinger-like equation of phonons is rewritten as $H_{\mathbf{k}}\psi_{\mathbf{k}} = \omega_{\mathbf{k}}\psi_{\mathbf{k}}$, with

$$H_{\mathbf{k}} = \begin{pmatrix} 0 & \mathrm{i}D_{\mathbf{k}}^{1/2} \\ -\mathrm{i}D_{\mathbf{k}}^{1/2} & -2\mathrm{i}\beta \end{pmatrix}, \quad \psi_{\mathbf{k}} = \begin{pmatrix} D_{\mathbf{k}}^{1/2}\epsilon_{\mathbf{k}} \\ -\mathrm{i}\omega_{\mathbf{k}}\epsilon_{\mathbf{k}} \end{pmatrix}, \tag{4.30}$$

$$D_{\mathbf{k}} = \sum_{l'} \Phi_{ll'} e^{\mathrm{i}\mathbf{k}\cdot(\mathbf{R}_{l'} - \mathbf{R}_l)}, \tag{4.31}$$

where $D_{\mathbf{k}}$ is the dynamical matrix. Here the dynamical matrix $D_{\mathbf{k}}$ is semi-positive-definite as required by the structural stability, and its square root $D_{\mathbf{k}}^{1/2}$ is thus Hermitian. $\beta$ is anti-symmetric tensor and $-2\mathrm{i}\beta$ is thus Hermitian. Therefore, the Hamiltonian $H_{\mathbf{k}}$ is Hermitian.

## 4.3 Magnetoelectric Effect for Phonons

In this section, we discuss generation of phonon angular momentum by an electric field. We calculate the phonon angular momentum using a toy model which is a two-dimensional spring-mass model with localized spins, and this model is shown in Fig. 4.1a. In Fig. 4.1a, the red circles with filled circles and crosses represent particles with spins perpendicular to $xy$ plane and effective negative charges, and the blue circles represent the particles with effective positive charges. This model has three particles in the unit cell and it is electrically and magnetically neutral. The

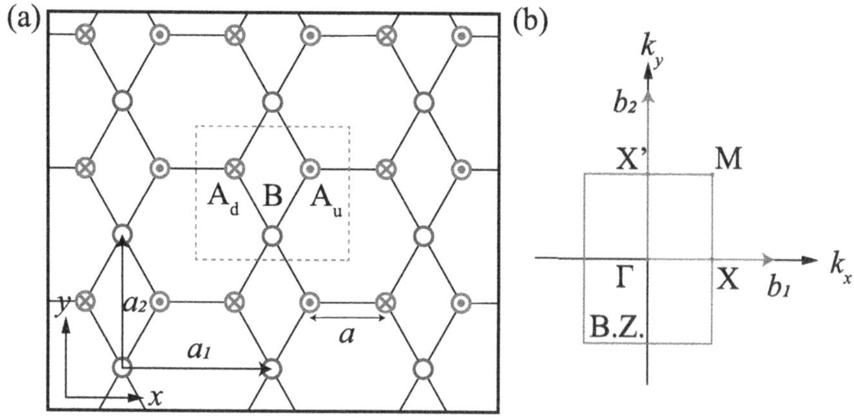


**Fig. 4.1** Model for generation of a phonon angular momentum by an electric field. **a** Spring-mass model with localized spins, and **b** its first Brillouin zone. The red circles with filled circles ($A_u$) and crosses ($A_d$) represent particles with spins perpendicular to $x$-$y$ plane and effective negative charges, and the blue circles (B) represent particles with effective positive charges. The primitive vectors are $\mathbf{a}_1 = a(2, 0)$, $\mathbf{a}_2 = a(0, \sqrt{3})$ with lattice constant $a$. Reprinted figure with permission from [2] Masato Hamada and Shuichi Murakami, "Phonon rotoelectric effect", Phys. Rev. B 101, 144306, 2020, Copyright 2020 by the American Physical Society

motions of the particles are confined within the $xy$ plane. Thus, this model represents an antiferromagnetic insulator. Here, the primitive vectors are $\mathbf{a}_1 = a(2, 0)$, $\mathbf{a}_2 = a(0, \sqrt{3})$ with lattice constant $a$, and the reciprocal vectors are $\mathbf{b}_1 = (2\pi/a)(1/2, 0)$, $\mathbf{b}_2 = (2\pi/a)(0, 1/\sqrt{3})$ and the first Brillouin zone is shown in Fig. 4.1b. We label the particles as shown as the red circles with filled circles, those with crosses and blue circles as $A_u$, $A_d$, and B, respectively. The spring constants of $A_u$-$A_d$ bonds and A-B bonds are $k_1$ and $k_2$, respectively. For simplicity, we assume that the Coulomb potential due to the effective charges is included in the spring constant. The potential energy of the spring is represented as

$$U = \frac{1}{2} \sum_{\langle |l\kappa, l'\kappa' \rangle} k_{l\kappa, l'\kappa'} (|\mathbf{R}_{l\kappa} - \mathbf{R}_{l'\kappa'} + \mathbf{u}_{l\kappa} - \mathbf{u}_{l'\kappa'}| - l_0)^2, \tag{4.32}$$

where $l_0$ is the length of the springs in equilibrium, $k_{l\kappa,l'\kappa'}$ is the spring constant between the $\kappa$th particle in the $l$th unit cell and the $\kappa'$ particle in the $l'$ unit cell, and $\langle |l\kappa, l'\kappa' \rangle$ represents pairs for the nearest neighbor bonds. The force onto each particle at its equilibrium position is defined as

$$F_\alpha(l\kappa) = \left. \frac{\partial U}{\partial u_{l\kappa,\alpha}} \right|_0 . \tag{4.33}$$

The force constant matrix is defined as

$$\Phi_{\alpha\beta}(l\kappa, l'\kappa') = \left.\frac{\partial^2 U}{\partial u_{l\kappa,\alpha}\partial u_{l'\kappa',\beta}}\right|_0. \tag{4.34}$$

Here, the nearest neighbor bond vectors are given by $\mathbf{d}_1 = a(1, 0)$, $\mathbf{d}_2 = a(-1/2, \sqrt{3}/2)$, $\mathbf{d}_3 = a(1/2, \sqrt{3}/2)$, and the displacement vectors of the individual particles are written as $\mathbf{u}_{\mathrm{A_u}}, \mathbf{u}_{\mathrm{A_d}}, \mathbf{u}_{\mathrm{B}}$. First, we check the balance among the forces onto each particle at its equilibrium position. The force onto each particle at its equilibrium position is obtained as

$$\begin{aligned} F_\alpha(\mathrm{A_u}) &= -k_1(|\mathbf{d}|_1 - l_0)\frac{d_1^\alpha}{|\mathbf{d}_1|} + k_2(|\mathbf{d}_2| - l_0)\frac{d_2^\alpha}{|\mathbf{d}_2|} - k_2(|\mathbf{d}_3| - l_0)\frac{d_3^\alpha}{|\mathbf{d}_3|} \\ &= ((-k_1 + k_2)a(1-\eta), 0), \end{aligned} \tag{4.35}$$

$$\begin{aligned} F_\alpha(\mathrm{A_d}) &= k_1(|\mathbf{d}|_1 - l_0)\frac{d_1^\alpha}{|\mathbf{d}_1|} - k_2(|\mathbf{d}_2| - l_0)\frac{d_2^\alpha}{|\mathbf{d}_2|} + k_2(|\mathbf{d}_3| - l_0)\frac{d_3^\alpha}{|\mathbf{d}_3|} \\ &= ((k_1 - k_2)a(1-\eta), 0), \end{aligned} \tag{4.36}$$

$$\begin{aligned} F_\alpha(\mathrm{B}) &= k_2(|\mathbf{d}|_2 - l_0)\frac{d_2^\alpha - d_2^\alpha}{|\mathbf{d}_2|} + k_2(|\mathbf{d}_3| - l_0)\frac{d_3^\alpha - d_3^\alpha}{|\mathbf{d}_3|} \\ &= (0, 0), \end{aligned} \tag{4.37}$$

with $\eta = l_0/a$. To make the force onto each particle at its equilibrium position zero, the spring constants $k_1, k_2$ are equal, and we set $k_1 = k_2 = k$. The force constant matrix is represented as

$$\begin{aligned} \Phi_{\alpha\beta}(l\kappa, l'\kappa') &= \left.\frac{\partial^2 U}{\partial u_\alpha(l\kappa)\partial u_\beta(l'\kappa')}\right|_0 \\ &= -k\frac{d_a^\beta d_a^\alpha}{|d_a|^2} - k(|d_a| - l_0)\frac{\delta_{\alpha\beta}|d_a| - d_a^\alpha \frac{d_a^\beta}{|d_a|}}{|d_a|^2} \\ &= -k\eta\frac{d_a^\alpha d_a^\beta}{|d_a|^2} - k(1-\eta)\delta_{\alpha\beta}. \end{aligned} \tag{4.38}$$

Then, the force constant matrices are written as

$$\Phi(\mathrm{A_u}, \mathrm{A_d}) = \begin{pmatrix} -k & 0 \\ 0 & -k(1-\eta) \end{pmatrix}, \tag{4.39}$$

$$\Phi(\mathrm{A_u}, \mathrm{B}) = \begin{pmatrix} -\frac{k}{4}(1-3\eta) & \frac{\sqrt{3}k}{4}(2-\eta) \\ \frac{\sqrt{3}k}{4}(2-\eta) & -\frac{k}{4}(1-\eta) \end{pmatrix}, \tag{4.40}$$

$$\Phi(\mathrm{A_d}, \mathrm{B}) = \begin{pmatrix} -\frac{k}{4}(1-3\eta) & -\frac{\sqrt{3}k}{4}(2-\eta) \\ -\frac{\sqrt{3}k}{4}(2-\eta) & -\frac{k}{4}(1-\eta) \end{pmatrix}. \tag{4.41}$$

In addition, because of infinitesimal translational invariance of the crystals, the on-site force constant matrix satisfies

$$\Phi_{\alpha\beta}(l\kappa, l\kappa) = -\sum_{l'\kappa' \neq l\kappa} \Phi_{\alpha\beta}(l\kappa, l'\kappa'), \tag{4.42}$$

and the on-site force constant matrices $\Phi(A, A)$ and $\Phi(B, B)$ are written as

$$\Phi(\mathrm{A}, \mathrm{A}) = \begin{pmatrix} k + \frac{k}{2}(1-\eta) & 0 \\ 0 & k(1-\eta) - \frac{k}{2}(1-\eta) \end{pmatrix}, \tag{4.43}$$

$$\Phi(\mathrm{B}, \mathrm{B}) = \begin{pmatrix} k(1-3\eta) & 0 \\ 0 & k(1-\eta) \end{pmatrix}. \tag{4.44}$$

Therefore, the dynamical matrix $D$ is obtained as

$$D = \begin{pmatrix} D(\mathrm{A_u}, \mathrm{A_u}) & D(\mathrm{A_u}, \mathrm{A_d}) & D(\mathrm{A_u}, \mathrm{B}) \\ D(\mathrm{A_d}, \mathrm{A_u}) & D(\mathrm{A_d}, \mathrm{A_d}) & D(\mathrm{A_d}, \mathrm{B}) \\ D(\mathrm{B}, \mathrm{A_u}) & D(\mathrm{B}, \mathrm{A_d}) & D(\mathrm{B}, \mathrm{B}) \end{pmatrix}, \tag{4.45}$$

$$D(\kappa, \kappa') = \sum_{l'} \frac{\Phi(l\kappa, l'\kappa')}{\sqrt{m_\alpha m_\beta}} e^{i(\mathbf{R}_{l'} - \mathbf{R}_l)\cdot \mathbf{k}}. \tag{4.46}$$

The Hamiltonian for the phonon in this toy model is represented as

$$H = \sum_l \left[ \frac{1}{2}(p_l - \beta_z u_l)^T (p_l - \beta_z u_l) + \frac{1}{2} \sum_{l'} u_l^T \Phi_{ll'} u_{l'} \right], \tag{4.47}$$

$$\beta_z = g \begin{pmatrix} S & 0 & 0 \\ 0 & -S & 0 \\ 0 & 0 & 0 \end{pmatrix} \otimes \begin{pmatrix} 0 & -1 \\ 1 & 0 \end{pmatrix}, \tag{4.48}$$

where $S$ is the spin of the A particles, and $g$ is a coupling constant of the spin-phonon interaction.

Next, we consider the phonon angular momentum generated by an electric field. First, we discuss the response tensor between the phonon angular momentum and the electric field. Next, we calculate the phonon angular momentum in this toy model.

### 4.3.1 Form of the Response Tensor Using the Magnetic Point Group in This Toy Model

Here, we discuss the response tensor between the phonon angular momentum and the electric field. The symmetry of this toy model is characterized by a magnetic point group, which is a symmetry including the time-reversal symmetry. The symmetry operations in this toy model are

$$\{E, \sigma_z, C_{2y}, \sigma_x\} + \mathrm{TR} \times \{I, C_{2z}, \sigma_y, C_{2x}\}, \tag{4.49}$$

where $E$ is the identity operator, $\sigma_z$ is the mirror reflection with respect to the $xy$ plane, $C_{2y}$ is the two-fold rotation around the $y$ axis, $\sigma_x$ is the mirror reflection with respect to the $yz$ plane, $I$ is the inversion operator, $C_{2z}$ is the two-fold rotation around the $z$ axis, $\sigma_y$ is the mirror reflection with respect to the $zx$ plane, $C_{2x}$ is the two-fold rotation around the $x$ axis, and TR is the time-reversal operation. We show symmetry properties of the phonon angular momentum using the magnetic point group. The generators of this magnetic point group are $\sigma_z, C_{2y}, \mathrm{TR} \times I$. In equilibrium, the phonon angular momentum for each mode becomes zero due to $\mathrm{TR} \times I$ symmetry, and therefore the total phonon angular momentum is zero. We consider the response tensor $\alpha$ of the phonon angular momentum when an electric field is applied. The response tensor is an axial tensor, and the nonzero elements of response tensor are determined only by the magnetic point-group symmetry. By using the generators of this point group, the form of the response tensor is determined as $\alpha = (-1)\sigma_z \alpha \sigma_z^{-1}$, $\alpha = C_{2y} \alpha C_{2y}^{-1}$, and $\alpha = (\mathrm{TR} \times I)\alpha(\mathrm{TR} \times I)^{-1}$, respectively. Namely, the three generators $\sigma_z$, $C_{2y}$ and $\mathrm{TR} \times I$ restrict the form of the tensor as

$$\alpha = \begin{pmatrix} 0 & 0 & \alpha_{xz} \\ 0 & 0 & \alpha_{yz} \\ \alpha_{zx} & \alpha_{zy} & 0 \end{pmatrix}, \alpha = \begin{pmatrix} \alpha_{xx} & 0 & \alpha_{xz} \\ 0 & \alpha_{yy} & 0 \\ \alpha_{zx} & 0 & \alpha_{zz} \end{pmatrix}, \alpha = \begin{pmatrix} \alpha_{xx} & \alpha_{xy} & \alpha_{xz} \\ \alpha_{yx} & \alpha_{yy} & \alpha_{yz} \\ \alpha_{zx} & \alpha_{zy} & \alpha_{zz} \end{pmatrix}, \tag{4.50}$$

respectively. Therefore, the response tensor is represented as

$$\alpha = \begin{pmatrix} 0 & 0 & \alpha_{xz} \\ 0 & 0 & 0 \\ \alpha_{zx} & 0 & 0 \end{pmatrix}. \tag{4.51}$$

This form of the response tensor $\alpha$ can also be seen from the following argument. When the electric field is applied along $x$ direction, the equilibrium positions of particles are slightly shifted along the $x$ direction due to the electrostatic force, and thus, the symmetry of this system is lowered to

$$\{E, \sigma_z\} + \mathrm{TR} \times \{\sigma_y, C_{2x}\}, \tag{4.52}$$

which allow an angular momentum along the $z$-direction. On the other hand, when the electric field is applied along $y$ direction, the equilibrium positions of particles are slightly shifted along the $y$ direction, and the symmetry of this system becomes

$$\{E, \sigma_z, C_{2y}, \sigma_x\}, \tag{4.53}$$

leading to cancellation of the phonon angular momentum along the $z$ direction between $(k_x, k_y)$ and $(-k_x, k_y)$. This analysis is consistent with the response tensor Eq. (4.51). Therefore, when the electric filed is applied along $x$ direction, the phonon angular momentum is generated and is proportional to electric field.

### 4.3.2 Numerical Results

We calculate the phonon angular momentum generated by the electric field. In this toy model, let $\tilde{\mathbf{d}}_\kappa$ denote the slight deviation of the particles along the external electric field $\mathbf{E}$. This lattice deformation comes from the electrostatic force onto the effective charges of the particles. In the deformed lattice structure, the new equilibrium positions become $\mathbf{R}'_{l\kappa} = \mathbf{R}_{l\kappa} + \tilde{\mathbf{d}}_\kappa$. The potential energy is rewritten as

$$U = \frac{1}{2} \sum_{\langle l\kappa, l'\kappa' \rangle} k(|\mathbf{R}'_{l\kappa} - \mathbf{R}'_{l'\kappa'} + \mathbf{u}_{l\kappa} - \mathbf{u}_{l'\kappa'}| - l_0)^2. \tag{4.54}$$

We derive the new force constant matrix $\Phi'$. Before deriving the new force constant matrix, we check the balance of the force onto each particle at its equilibrium position. From the definition of the force onto each particle at its equilibrium position Eq. (4.33), one has

$$\mathbf{F}(\mathrm{A_u}) = -k\left(1 - \frac{l}{|\mathbf{d}'_1|}\right)\mathbf{d}'_1 + k\left(1 - \frac{l}{|\mathbf{d}'_{21}|}\right)\mathbf{d}'_{21} - k\left(1 - \frac{l}{|\mathbf{d}'_{31}|}\right)\mathbf{d}'_{31}, \tag{4.55}$$

$$\mathbf{F}(\mathrm{A_d}) = k\left(1 - \frac{l}{|\mathbf{d}'_1|}\right)\mathbf{d}'_1 - k\left(1 - \frac{l}{|\mathbf{d}'_{22}|}\right)\mathbf{d}'_{22} + k\left(1 - \frac{l}{|\mathbf{d}'_{32}|}\right)\mathbf{d}'_{32}, \tag{4.56}$$

$$\begin{aligned}\mathbf{F}(\mathrm{B}) = &-k\left(1 - \frac{l}{|\mathbf{d}'_{21}|}\right)\mathbf{d}'_{21} + k\left(1 - \frac{l}{|\mathbf{d}'_{22}|}\right)\mathbf{d}'_{22} \\ &+ k\left(1 - \frac{l}{|\mathbf{d}'_{31}|}\right)\mathbf{d}'_{31}, -k\left(1 - \frac{l}{|\mathbf{d}'_{32}|}\right)\mathbf{d}'_{32},\end{aligned} \tag{4.57}$$

where, $\mathbf{d}'_1, \mathbf{d}'_{21}, \mathbf{d}'_{22}, \mathbf{d}'_{31}, \mathbf{d}'_{32}$ are the new nearest neighbor bond vectors. For example, when the electric field along $x$-direction is applied, the new nearest neighbor bond vectors are given by $\mathbf{d}'_1 = \mathbf{d}_1, \mathbf{d}'_{21} = \mathbf{d}_2 - \tilde{\mathbf{d}}, \mathbf{d}'_{22} = \mathbf{d}_2 + \tilde{\mathbf{d}}, \mathbf{d}'_{31} = \mathbf{d}_3 + \tilde{\mathbf{d}}$, and $\mathbf{d}'_{32} = \mathbf{d}_3 - \tilde{\mathbf{d}}$. Then, the force onto each particle at its equilibrium position is obtained as

$$\mathbf{F}(\mathrm{A_u}) = -k(1-\eta)\mathbf{d}_1 + k\left(1 - \frac{\eta}{1 + \frac{d}{2a}}\right)(\mathbf{d}_2 - \mathbf{d}_3 + 2\tilde{\mathbf{d}}), \tag{4.58}$$

$$\mathbf{F}(\mathrm{A_d}) = k(1-\eta)\mathbf{d}_1 - k\left(1 - \frac{\eta}{1 - \frac{d}{2a}}\right)(\mathbf{d}_2 - \mathbf{d}_3 - 2\tilde{\mathbf{d}}), \tag{4.59}$$

$$\mathbf{F}(\mathrm{B}) = -k\left(1 - \frac{\eta}{1+\frac{d}{2a}}\right)(\mathbf{d}_2 - \mathbf{d}_3 + 2\tilde{\mathbf{d}}) + k\left(1 - \frac{\eta}{1-\frac{d}{2a}}\right)(\mathbf{d}_2 - \mathbf{d}_3 - 2\tilde{\mathbf{d}}). \tag{4.60}$$

In addition, the electrostatic force work on each particle in the unit cell is given by $\mathbf{F}_{\mathrm{ex}}(\mathrm{A_u}) = \mathbf{F}_{\mathrm{ex}}(\mathrm{A_d}) = -q_0\mathbf{E}$ and $\mathbf{F}_{\mathrm{ex}}(\mathrm{B}) = 2q_0\mathbf{E}$ with effective charge $q_0$. To satisfy the balance of the force onto each particle at its equilibrium position, the slightly deviation $\tilde{\mathbf{d}} = (\tilde{d}, 0)$ is represented as

$$\tilde{d} = \frac{2q_0}{k(4-3\eta)}E_x, \tag{4.61}$$

which shows that the slight deviation is proportional to the electric field. Next, we derive the new force constant matrix. From Eq. (4.38), we get

$$\Phi(\mathrm{A_u}, \mathrm{A_d})' = \begin{pmatrix} -k & 0 \\ 0 & -k(1-\eta) \end{pmatrix}, \tag{4.62}$$

$$\Phi_{(21)}(\mathrm{A_u}, \mathrm{B})' = -\frac{k\eta}{r_-^2}\begin{pmatrix} (\frac{a}{2}-\tilde{d})^2 & \frac{\sqrt{3}}{2}a(\frac{a}{2}-\tilde{d}) \\ \frac{\sqrt{3}}{2}a(\frac{a}{2}-\tilde{d}) & -\frac{3}{4}a^2 \end{pmatrix} - k(1-\eta)\begin{pmatrix} 1 & 0 \\ 0 & 1 \end{pmatrix}, \tag{4.63}$$

$$\Phi_{(31)}(\mathrm{A_u}, \mathrm{B})' = -\frac{k\eta}{r_-^2}\begin{pmatrix} (\frac{a}{2}-\tilde{d})^2 & -\frac{\sqrt{3}}{2}a(\frac{a}{2}-\tilde{d}) \\ -\frac{\sqrt{3}}{2}a(\frac{a}{2}-\tilde{d}) & -\frac{3}{4}a^2 \end{pmatrix} - k(1-\eta)\begin{pmatrix} 1 & 0 \\ 0 & 1 \end{pmatrix}, \tag{4.64}$$

$$\Phi_{(22)}(\mathrm{A_d}, \mathrm{B})' = -\frac{k\eta}{r_+^2}\begin{pmatrix} (\frac{a}{2}+\tilde{d})^2 & \frac{\sqrt{3}}{2}a(\frac{a}{2}+\tilde{d}) \\ \frac{\sqrt{3}}{2}a(\frac{a}{2}+\tilde{d}) & -\frac{3}{4}a^2 \end{pmatrix} - k(1-\eta)\begin{pmatrix} 1 & 0 \\ 0 & 1 \end{pmatrix}, \tag{4.65}$$

$$\Phi_{(32)}(\mathrm{A_d}, \mathrm{B})' = -\frac{k\eta}{r_+^2}\begin{pmatrix} (\frac{a}{2}+\tilde{d})^2 & -\frac{\sqrt{3}}{2}a(\frac{a}{2}+\tilde{d}) \\ -\frac{\sqrt{3}}{2}a(\frac{a}{2}+\tilde{d}) & -\frac{3}{4}a^2 \end{pmatrix} - k(1-\eta)\begin{pmatrix} 1 & 0 \\ 0 & 1 \end{pmatrix} \tag{4.66}$$

with $r_\pm \simeq a \pm \tilde{d}$. In addition, the on-site force matrix constant is rewritten as

$$\Phi(\mathrm{A_u}, \mathrm{A_u})' = -\Phi(\mathrm{A_u}, \mathrm{A_d})' - \Phi_{(21)}(\mathrm{A_u}, \mathrm{B})' - \Phi_{(31)}(\mathrm{A_u}, \mathrm{B})', \tag{4.67}$$

$$\Phi(\mathrm{A_d}, \mathrm{A_d})' = -\Phi(\mathrm{A_u}, \mathrm{A_d})' - \Phi_{(22)}(\mathrm{A_d}, \mathrm{B})' - \Phi_{(32)}(\mathrm{A_d}, \mathrm{B})', \tag{4.68}$$

$$\Phi(\mathrm{B}, \mathrm{B})' = -\Phi_{(21)}(\mathrm{A_u}, \mathrm{B})' - \Phi_{(31)}(\mathrm{A_u}, \mathrm{B})' - \Phi_{(22)}(\mathrm{A_d}, \mathrm{B})' - \Phi_{(32)}(\mathrm{A_d}, \mathrm{B})'. \tag{4.69}$$

Therefore, the new dynamical matrix $D'$ is obtained from Eq. (4.46).

Here we show the result of our calculation. First, in Fig. 4.2a, we show the phonon dispersion without the electric field. In this model, there are six phonon modes because this model consists of three particles per unit cell and we consider two-dimensional motions. For simplicity, we set the parameters as $k = 1$ and $m_A = m_B = 1$. We introduce a parameter $\eta = l_0/a$ representing the restoring force. Note that if $\eta = 1$, this model is not stretched, and the restoring force does not work, while if $\eta < 1$, this system is stretched, and the restoring force works. When the restoring force does not work ($\eta = 1$), the phonon dispersion has zero frequency flat band and this model is unstable. Therefore we set $\eta < 1$. In Fig. 4.2b, we show the phonon angular momentum for each mode, meaning that the phonon angular momentum is zero at all **k**. This result is consistent with the conclusion from symmetry. In Fig. 4.2c, we show the phonon angular momentum of each particle for the second lowest band. The phonon angular momenta of $A_u$ and $A_d$ are nonzero with equal magnitude and opposite signs. The phonon angular momentum of B is always zero. In Fig. 4.2d, we show the trajectory of particles for the second lowest band at $X$ point. The Lorentz force due to the spin-phonon interaction is exerted onto the particles $A_u$ and $A_d$ with opposite directions, while this force is absent at the particle B. Therefore, the particles $A_u$ and $A_d$ rotate in opposite directions, while the particles $B$ vibrate but do not rotate.

Next, we show the phonon dispersion under the electric field $E_x$ in Fig. 4.3. In Fig. 4.3a, we show the deformed lattice structure. In Fig. 4.3b, we show the phonon dispersion with the electric field $E_x$. The phonon dispersion with the electric field $E_x$ is almost the same as that without electric field (Fig. 4.2a) because the deviation $\tilde{d}$ is small. We show the phonon angular momentum for each mode in Fig. 4.3c, and also show the phonon angular momentum of each particle for the second lowest band in Fig. 4.3d. In Fig. 4.3c, the phonon angular momentum for each mode has a nonzero value, and, in Fig. 4.3d, the phonon angular momentum of each particle is different from that without an electric field (Fig. 4.2c). We also show the trajectories of particles for the second lowest band at $X$ point in Fig. 4.3e. Since the symmetry is lowered and the $C_{2y}$ and $\sigma_x$ symmetries are broken, the phonon angular momentum of $A_u$ does not cancel that of $A_d$. Moreover, the phonon angular momentum of B can also become nonzero. In Fig. 4.4a and b, we show the sum of the phonon angular momentum of all modes with the electric field $E_x$ on the high-symmetry line and in the first Brillouin zone, respectively. The sum of the phonon angular momentum along the $z$ direction of all mode becomes nonzero. At finite temperature, the total phonon angular momentum is given by the sum of the product between $l_{\sigma,z}(\mathbf{k})$ and the Bose distribution function, and it is also nonzero. Therefore, the phonon angular momentum is generated by electric field.

Next, we also show the phonon dispersion under the electric filed $E_y$ in Fig. 4.5. In Fig. 4.5a, we show the lattice structure deformed along the $y$ direction. In Fig. 4.5b, we show the phonon dispersion with the electric field $E_y$. We show the phonon angular momentum for each mode and the phonon angular momentum of each particle for the second lowest band in Fig. 4.5c and d, respectively. The phonon angular momentum for each mode has nonzero value at wave vector **k** since the product of time-reversal and inversion symmetries TR $\times$ I is broken. Along the $X - M$ and $X' - \Gamma$ lines, the phonon angular momentum along the $z$ direction for each mode is zero since the $C_{2y}$ symmetry is conserved. Moreover, the phonon angular momentum of B has a nonzero value. In Fig. 4.5e, we show the trajectories of particles for the second lowest bands at $X$ point. We show the sum of the phonon angular momentum of all modes with the electric field $E_y$ on the high-symmetry line and in the first Brillouin zone in Fig. 4.6a and b, respectively. The sum of the phonon angular momentum of all modes

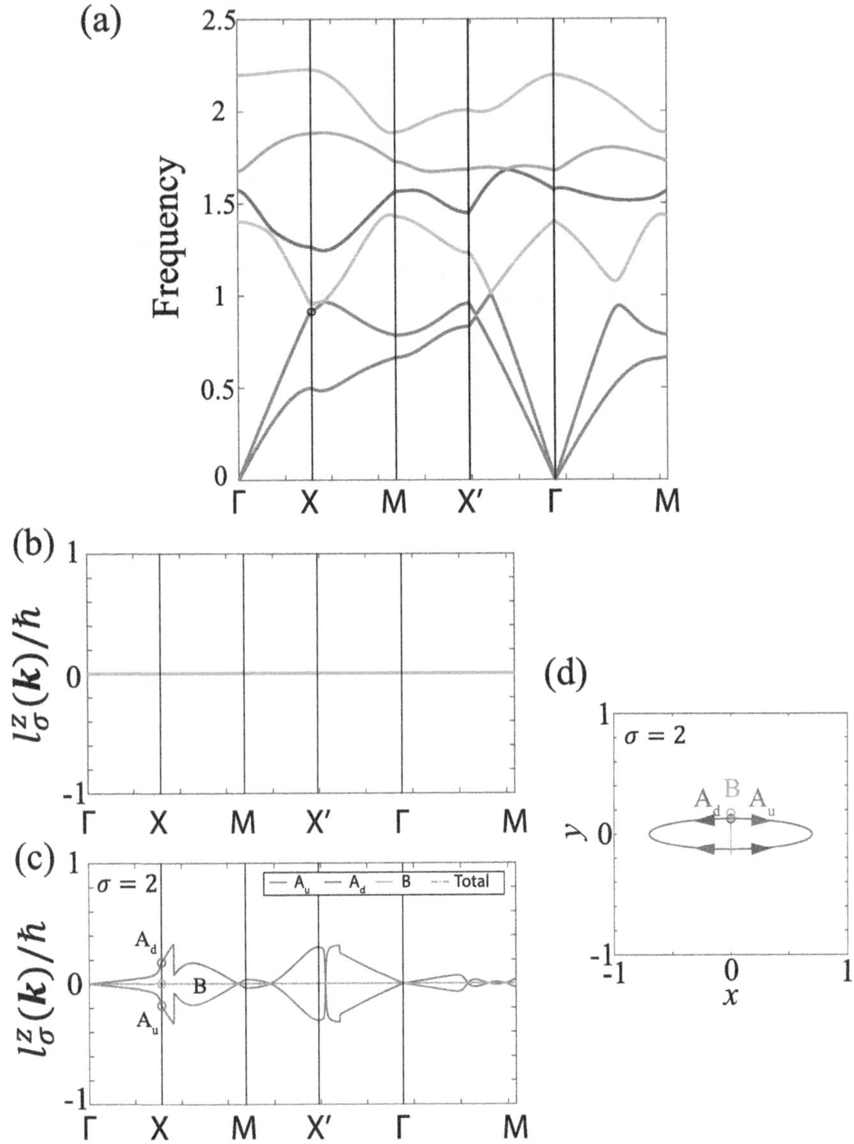


**Fig. 4.2** Phonon dispersion and the phonon angular momentum without an electric field. **a** Phonon dispersion without electric field. **b** Phonon angular momentum along $z$-direction for each mode without an electric field. **c** Phonon angular momentum of each particle and its sum for the second lowest band. **d** Trajectories of the three particles in the unit cell for the phonons in the second lowest band at $X$ point, which is indicated as circles in **c**. In **d** we show the normalized polarization vector $\epsilon_{\mathbf{k}\sigma}$ with $\epsilon^\dagger_{\mathbf{k}\sigma}\epsilon_{\mathbf{k}\sigma} + \frac{i}{\omega_{\mathbf{k}\sigma}}\epsilon_{\mathbf{k}\sigma}\beta_z\epsilon_{\mathbf{k}\sigma} = 1$, and their axes $(x, y)$ are shown in a dimensionless unit. We set the parameter as $k = 1$, $m_\mathrm{A} = m_\mathrm{B} = 1$, $\eta = 0.8$, and $\tilde{\mathbf{d}} = (0, 0)$

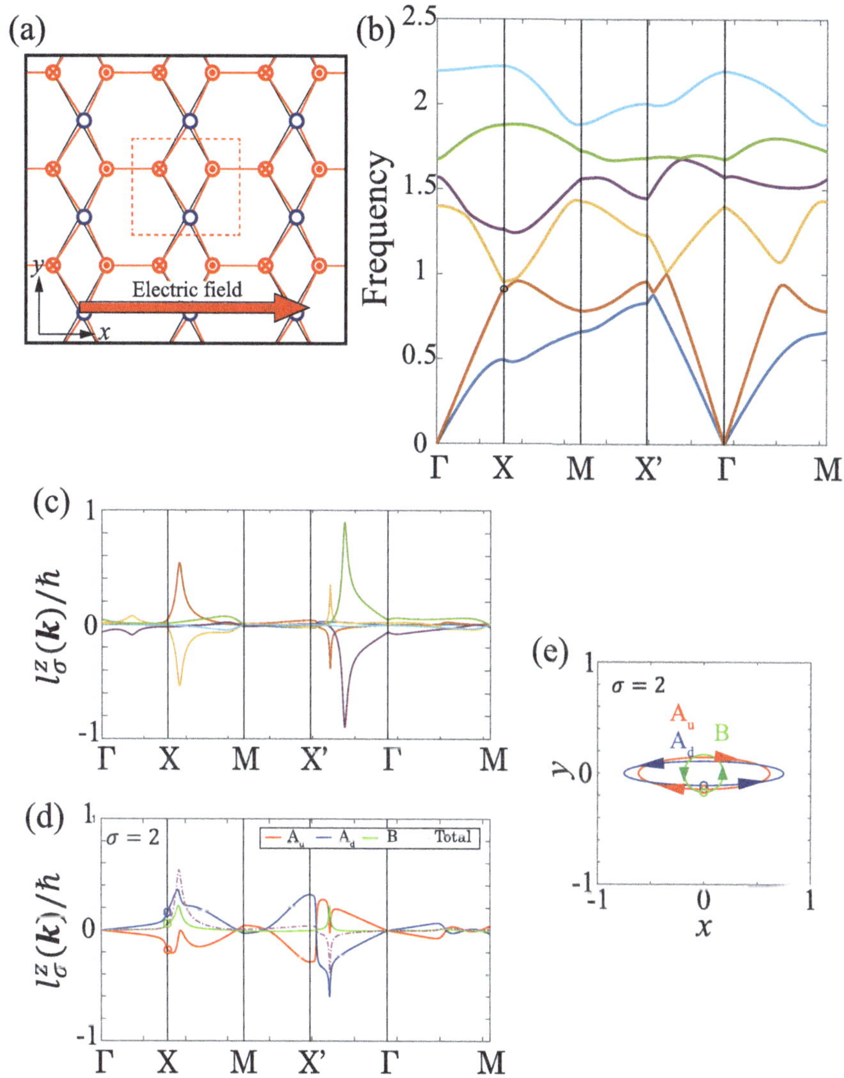


**Fig. 4.3** Phonon dispersion and the phonon angular momentum with the electric field along $x$-direction. **a** Phonon dispersion with the electric field $E_x$. **b** Phonon angular momentum along $z$-direction for each mode with the electric field $E_x$. **c** Phonon angular momentum of each particle and its sum for the second lowest band. **d** Trajectories of the three particles in the unit cell for the phonon in the second lowest band at $X$ point, which is indicated as circles in **c**. In **d** we show the normalized polarization vector $\epsilon_{\mathbf{k}\sigma}$ with $\epsilon^{\dagger}_{\mathbf{k}\sigma}\epsilon_{\mathbf{k}\sigma} + \frac{i}{\omega_{\mathbf{k}\sigma}}\epsilon_{\mathbf{k}\sigma}\beta_z\epsilon_{\mathbf{k}\sigma} = 1$, and their axes $(x, y)$ are shown in a dimensionless unit. We set the parameter as $k = 1$, $m_{\mathrm{A}} = m_{\mathrm{B}} = 1$, $\eta = 0.8$, and $\tilde{\mathbf{d}} = (a/50, 0)$. Reprinted figure with permission from [2] Masato Hamada and Shuichi Murakami, "Phonon rotoelectric effect", Phys. Rev. B 101, 144306, 2020, Copyright 2020 by the American Physical Society

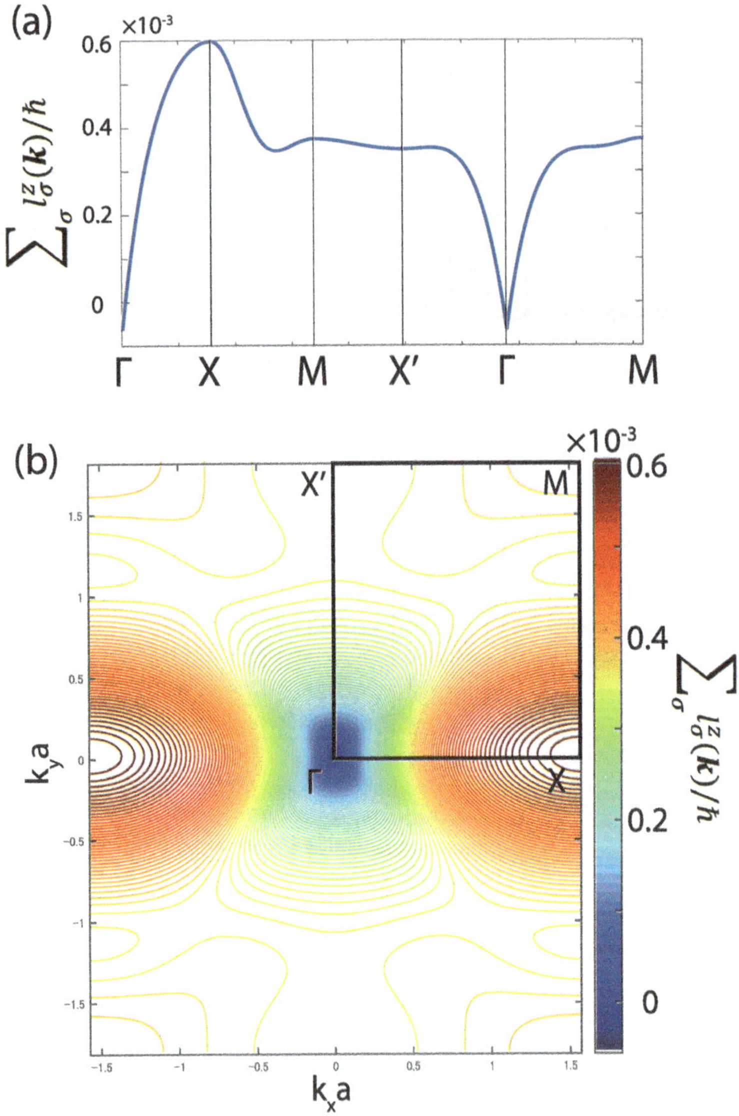


**Fig. 4.4** Sum of the phonon angular momentum $\sum_\sigma l_{\mathbf{k}\sigma,z}$ under an electric field $E_x$. **a** Sum of the phonon angular momentum with an electric field $E_x$ on the high symmetry lines. **b** Sum of the phonon angular momentum with an electric field $E_x$, shown in the first Brillouin zone. Modified figure with permission from [2] Masato Hamada and Shuichi Murakami, "Phonon rotoelectric effect", Phys. Rev. B 101, 144306, 2020, Copyright 2020 by the American Physical Society

cancels between $(k_x, k_y)$ and $(-k_x, k_y)$ due to $C_{2y}$ and $\sigma_x$ symmetries. Therefore, the phonon angular momentum with the electric field $E_y$ have the nonzero value at any wave vector **k**, and its sum over the wave vector in the first Brillouin zone vanishes.

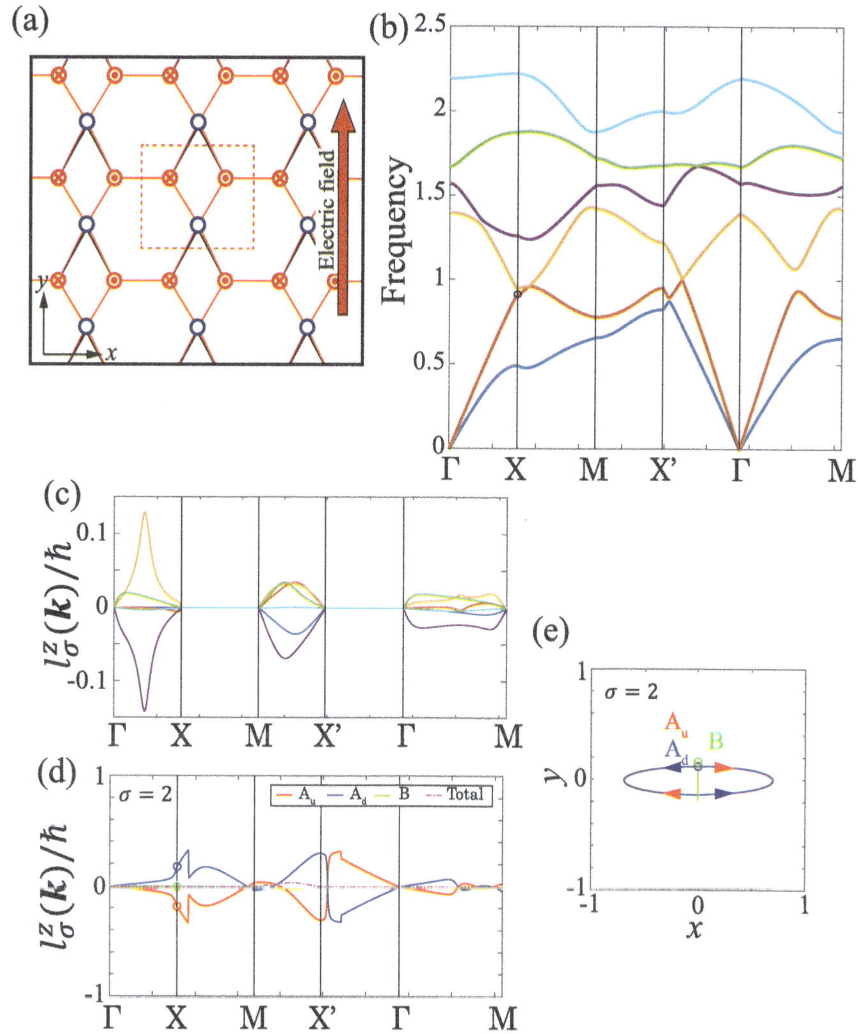


**Fig. 4.5** Phonon dispersion and the phonon angular momentum with the electric field along $y$-direction. **a** Phonon dispersion with the electric field $E_y$. **b** Phonon angular momentum along $z$-direction for each mode with an electric field $E_y$. **c** Phonon angular momentum of each particle and its sum for the second lowest band. **d** Trajectories of the three particles in the unit cell for the phonon in the second lowest band at $X$ point, which is indicated as circles in **c**. In **d** we show the normalized polarization vector $\epsilon_{\mathbf{k}\sigma}$ with $\epsilon^{\dagger}_{\mathbf{k}\sigma}\epsilon_{\mathbf{k}\sigma} + \frac{i}{\omega_{\mathbf{k}\sigma}}\epsilon_{\mathbf{k}\sigma}\beta_z\epsilon_{\mathbf{k}\sigma} = 1$, and their axes $(x, y)$ are shown in a dimensionless unit. We set the parameter as $k = 1$, $m_{\mathrm{A}} = m_{\mathrm{B}} = 1$, $\eta = 0.8$, and $\tilde{\mathbf{d}} = (0, a/50)$. Reprinted figure with permission from [2] Masato Hamada and Shuichi Murakami, "Phonon rotoelectric effect", Phys. Rev. B 101, 144306, 2020, Copyright 2020 by the American Physical Society

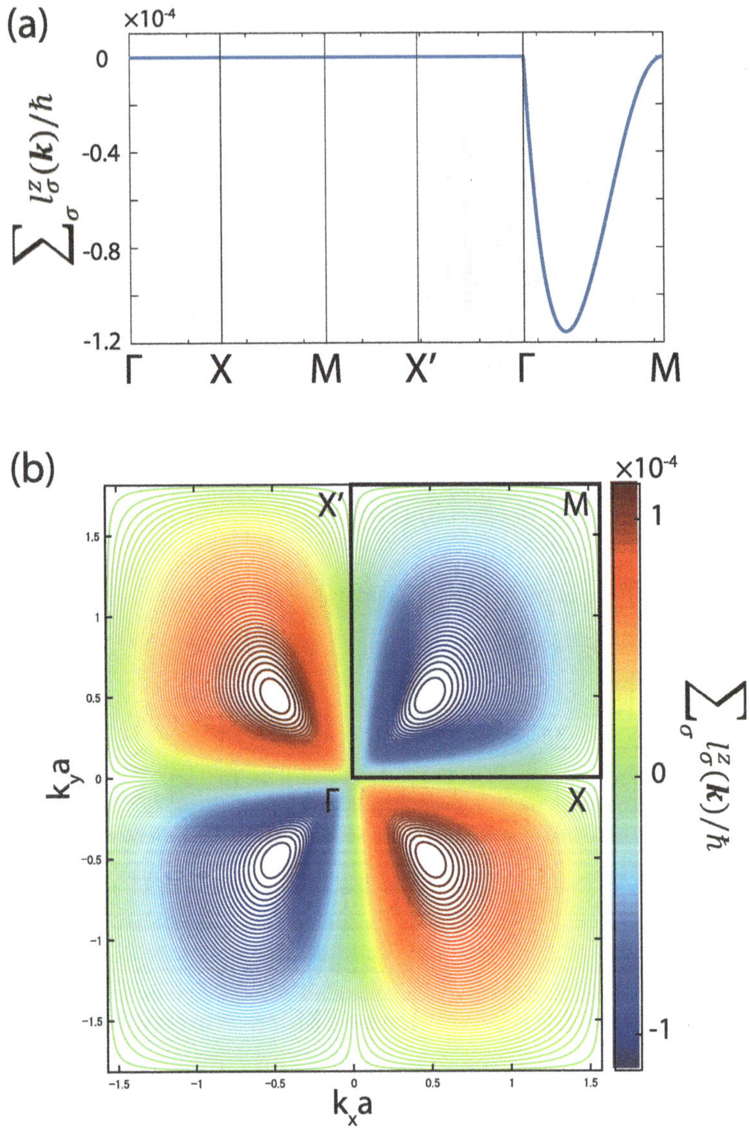


**Fig. 4.6** Sum of the phonon angular momentum $\sum_{\sigma} l_{\mathbf{k}\sigma,z}$ under an electric field $E_y$. **a** Sum of the phonon angular momentum with an electric field $E_y$ on the high symmetry lines. **b** Sum of the phonon angular momentum with an electric field $E_y$, shown in the first Brillouin zone. Modified figure with permission from [2] Masato Hamada and Shuichi Murakami, "Phonon rotoelectric effect", Phys. Rev. B 101, 144306, 2020, Copyright 2020 by the American Physical Society

## 4.4 Phonon Angular Momentum Due to the Magnetoelectric Effect in the High- and the Low-Temperature Limits

We have found that the phonon angular momentum is generated by the electric field. Here, we discuss temperature dependence of the phonon angular momentum generated by the electric field. The phonon angular momentum generated by the electric field is represented as Eq. (4.2). Using our toy model, we show the phonon angular momentum for the electric field $E_x$ in Fig. 4.7. In Fig. 4.7b, the sum of the phonon angular momentum of all modes over the wave vector in the first Brillouin zone is proportional to the slight deviation $\tilde{d}$. Because, from Eq. (4.61), the slight deviation is proportional to the electric field, the phonon angular momentum is proportional to the electric field. Indeed, in Fig. 4.7a, the sum of the phonon angular momentum for each mode over the wave vector is almost proportional to the electric field since the deviation $\tilde{d}$ is sufficiently small. The phonon angular momentum is represented as

$$J_z^{\mathrm{ph}} = \sum_{\mathbf{k}\sigma} l_{\mathbf{k}\sigma}^z \left( f_0(\omega_{\mathbf{k}\sigma}) + \frac{1}{2} \right), \tag{4.70}$$

and, in the system with an electric field, the phonon angular momentum for each mode is modulated by the lattice deformation. As in the previous study [10], in the high-temperature limit, the phonon angular momentum generated by an electric field vanishes. On the other hand, in the low-temperature limit, the Bose distribution function becomes negligible small, and the zero-point motion is dominant for the phonon

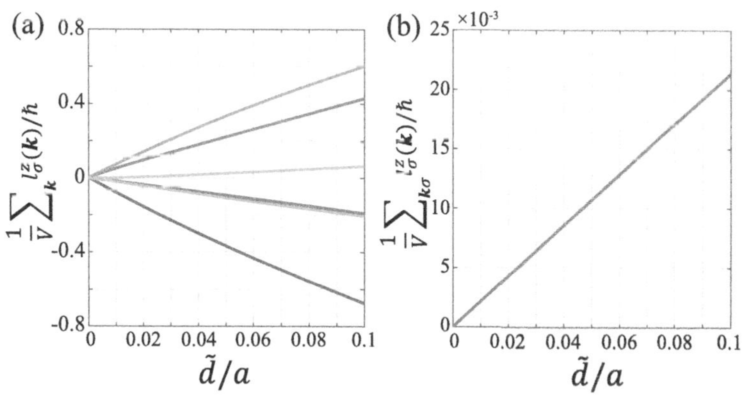


**Fig. 4.7** Dependence of the phonon angular momentum under the electric field $E_x$ on the atomic deviation $\tilde{d}$. **a** Phonon angular momentum for each mode. **b** Sum of the phonon angular momentum of all the modes. Reprinted figure with permission from [2] Masato Hamada and Shuichi Murakami, "Phonon rotoelectric effect", Phys. Rev. B 101, 144306, 2020, Copyright 2020 by the American Physical Society

angular momentum. To summarize, for the phonon angular momentum generated by an electric field, the zero-point motion of phonons is dominant and the resulting phonon angular momentum is proportional to the electric field in the low-temperature limit.

## 4.5 Summary

In summary, we have theoretically predicted generation of the phonon angular momentum in systems without time-reversal and inversion symmetries. When the temperature gradient is applied to systems with neither of these symmetries, the phonon angular momentum consists of two terms: an equilibrium term and a term proportional to the temperature gradient. For the phonon thermal Edelstein effect in magnetic crystals, the response tensor is determined by a magnetic point group. On the other hand, when both symmetries are broken and their product is conserved, the phonon angular momentum is generated by an electric field. This mechanism is analogous to the magnetoelectric effect in multiferroics materials, and we call it the phonon rotoelectric effect. We predict the phonon angular momentum is generated by the electric field using the magnetic point group, and we calculate the phonon angular momentum using the toy model. For the phonon angular momentum generated by an electric field, the modulation of the phonon angular momentum for each mode due to the lattice deformation is important, and temperature dependence is same as the phonon angular momentum in a magnetic field. Therefore, the phonon angular momentum due to the phonon rotoelectric effect vanish in the high-temperature limit. On the other hand, in the low-temperature limit, the zero-point motion of phonon angular momentum is dominant and is proportional to the electric field.

## References

1. Dzyaloshinskii IE (1959) On the magneto-electrical effect in antiferromagnets 10:628
2. Hamada M, Murakami S (2020) Phonon rotoelectric effect. Phys Rev B 101:144306. https://doi.org/10.1103/PhysRevB.101.144306
3. de L Kronig R (1939) On the mechanism of paramagnetic relaxation. Physica 6:33–43
4. Van Vleck JH (1940) Paramagnetic relaxation times for titanium and chrome alum. Phys Rev 57:426–447
5. Capellmann H, Lipinski S (1991) Spin-phonon coupling in intermediate valency: exactly solvable models. Zeitschrift für Physik B Condens Matter 83:199–205
6. Capellmann H, Neumann KU (1987) Intermediate valency i: couplings, hamiltonian, and electrical resistivity. Zeitschrift für Physik B Condens Matter 67:53–61
7. Capellmann H, Lipinski S, Neumann K-U (1989) A microscopic model for the coupling of spin fluctuations and charge fluctuation in intermediate valency. Zeitschrift fur Physik B Condens Matter 75:323–329
8. Birss RR (1962) In: Wohlfarth EP (ed) Symmetry and magnetism. Series of monographs on selected topics in solid state physics. Elsevier North-Holland
9. Liu Y, Xu Y, Zhang S-C, Duan W (2017) Model for topological phononics and phonon diode. Phys Rev B 96:064106
10. Zhang L, Niu Q (2014) Angular momentum of phonons and the einstein-de haas effect. Phys Rev Lett 112:085503

# Chapter 5
# Conversion Between Spins and Mechanical Rotations

## 5.1 Conversion Between Spins and Mechanical Rotations

In the Einstein-de Haas effect [1] and the Barnett effect [2], magnetization and mechanical rotation of a whole crystal are mutually converted. The origin of spin conversion in these effects is considered to be the spin-rotation coupling [3–6], which couples spin and a mechanical rotation. In spintronics, the mechanical generation of spin current via the spin-rotation coupling has been reported in various setups, such as a surface acoustic wave [7, 8], a twisting vibrational mode in a carbon nanotube [9], and a flow of a liquid metal [10]. Thus, the conversion between a spin and a mechanical rotation has been studied intensively. However, microscopic origins of the spin-rotation coupling are not well-known. The known form for the spin-rotation coupling was derived from a change to a rotational frame, and it might correspond to a rotational of the whole crystal but not to a microscopic local rotation of the lattice. Therefore, we focus on the conversion between a electron spin and a microscopic local rotation of atoms in a lattice, and we show that the electron spin is polarized along the microscopic local rotational axis [11].

## 5.2 Honeycomb-Lattice Model with the Periodic Microscopic Local Rotation

We introduce the microscopic local rotation in the honeycomb lattice. In the honeycomb lattice without inversion symmetry, there are four phonon bands for the atomic motions within the $xy$ plane and the phonon angular momentum for each mode becomes an odd function of $\mathbf{k}$ [12]. While the total phonon angular momentum becomes zero, each atom undergoes a rotational motion for each phonon band at any wave vector $\mathbf{k}$. Here, as an example, we focus on the rotational motion with the atoms A and B in the sublattice rotating in the same direction with their phases shifted by $\pi$ as shown in Fig. 5.1. This rotational motion corresponds to the atomic



M. Hamada, *Theory of Generation and Conversion of Phonon Angular Momentum*, Springer Theses, https://doi.org/10.1007/978-981-33-4690-1_5

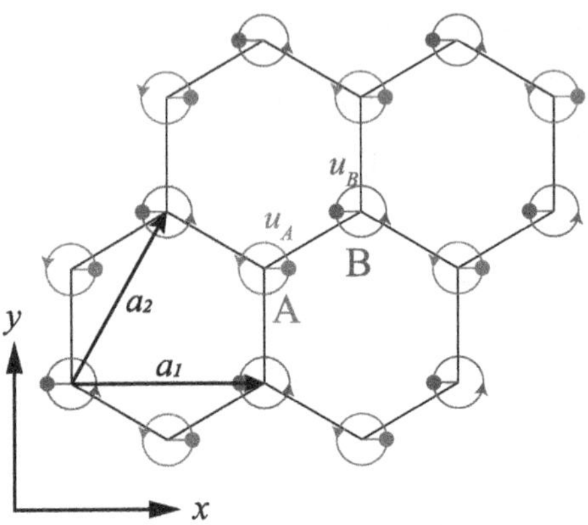


**Fig. 5.1** Schematic figure of the honeycomb lattice with periodic microscopic local rotations. Reprinted figure with permission from [11] Masato Hamada and Shuichi Murakami, "Conversion between electron spin and microscopic atomic rotation", Phys. Rev. Research 2. 023275, 2020, by the Author(s) licensed under CC BY 4.0

motion of the two degenerated optical phonon modes at the wave vector $\mathbf{k} = 0$. The primitive vectors are written as $\mathbf{a}_1 = a(1, 0)$, $\mathbf{a}_2 = a(1/2, \sqrt{3}/2)$ with a lattice constant $a$, and the vectors between the nearest neighbor atoms are written as $\mathbf{d}_1 = a_0(\sqrt{3}/2, 1/2)$, $\mathbf{d}_2 = a_0(-\sqrt{3}/2, 1/2)$, $\mathbf{d}_3 = a_0(0, -1)$ with the nearest bond length $a_0 = a/\sqrt{3}$. We set the angular velocity of the atoms A and B to be $\Omega$, and the displacement vector of each sublattice is represented as

$$\mathbf{u}_\mathrm{A} = u_0^\mathrm{A}(\cos\Omega t, \sin\Omega t), \ \ \mathbf{u}_\mathrm{B} = -u_0^B(\cos\Omega t, \sin\Omega t), \tag{5.1}$$

and the displacement vector from the atom A to the atom B is

$$\mathbf{u} = \mathbf{u}_\mathrm{B} - \mathbf{u}_\mathrm{A} = -u_+(\cos\Omega t, \sin\Omega t) \tag{5.2}$$

with $u_+ = u_0^\mathrm{A} + u_0^\mathrm{B}$.

Next, we consider the Hamiltonian for electrons in our toy model. First, we consider the tight-binding Hamiltonian without periodic microscopic local rotation, which is similar to the Kane-Mele model, representing a two-dimensional topological insulator [13, 14]:

$$H_0 = t_0 \sum_{\langle ij \rangle} c_i^\dagger c_j + \lambda_v \sum_i \xi_i c_i^\dagger c_i + \mathrm{i}\frac{\lambda_\mathrm{R}}{a_0} \sum_{\langle ij \rangle} c_i^\dagger (\mathbf{s} \times \mathbf{d}_{ij})_z c_j. \tag{5.3}$$

The first term is a nearest neighbor hopping term and $t_0$ is the hopping parameter. The second term is a staggered sublattice potential, which we include to break the inversion symmetry. $\xi_i$ is $\xi_{\mathrm{A(B)}} = +1(-1)$ for the atoms A and B and $\lambda_v$ is the on-site potential. The third term is a nearest neighbor Rashba term, which leads to spin-split band structure, and $\lambda_\mathrm{R}$ is Rashba parameter. The operators $c_i^\dagger = (c_{i\uparrow}^\dagger, c_{i\downarrow}^\dagger)$, $c_i = (c_{i\uparrow}, c_{i\downarrow})^T$ show the creation and annihilation operators for electron at the $i$ site, respectively. We introduce the Fourier transformation of the creation and annihilation operators

$$c_i = \frac{1}{\sqrt{N}} \sum_{\mathbf{k}} e^{\mathrm{i}\mathbf{k}\cdot\mathbf{r}_i} c_{\mathbf{k}},\ c_i^\dagger = \frac{1}{\sqrt{N}} \sum_{\mathbf{k}} e^{-\mathrm{i}\mathbf{k}\cdot\mathbf{r}_i} c_{\mathbf{k}}^\dagger. \tag{5.4}$$

The first term in Eq. (5.3) is rewritten as

$$\begin{aligned}(\text{the first term}) &= \frac{1}{N} \sum_{\langle|ij\rangle} \sum_{\mathbf{k},\mathbf{k}'} e^{-\mathrm{i}\mathbf{k}\cdot\mathbf{r}_i + \mathrm{i}\mathbf{k}'\cdot\mathbf{r}_j} c_{A\mathbf{k}}^\dagger c_{B\mathbf{k}} + e^{-\mathrm{i}\mathbf{k}\cdot\mathbf{r}_j + \mathrm{i}\mathbf{k}'\cdot\mathbf{r}_i} c_{B\mathbf{k}}^\dagger c_{A\mathbf{k}'} \\ &= \sum_{\mathbf{k}} \sum_{a=1,2,3} e^{\mathrm{i}\mathbf{k}\cdot\mathbf{d}_a} c_{A\mathbf{k}}^\dagger c_{B\mathbf{k}} + e^{-\mathrm{i}\mathbf{k}\cdot\mathbf{d}_a} c_{B\mathbf{k}}^\dagger c_{A\mathbf{k}},\end{aligned} \tag{5.5}$$

and the second term is rewritten as

$$(\text{the second term}) = \sum_{\mathbf{k}} \lambda_v c_{A\mathbf{k}}^\dagger c_{A\mathbf{k}} - \lambda_v c_{B\mathbf{k}}^\dagger c_{B\mathbf{k}}. \tag{5.6}$$

Next, we rewrite the Rashba term $(\mathbf{s} \times \mathbf{d}_a)_z$ as

$$(\mathbf{s} \times \mathbf{d}_{a=1,2,3})_z = a_0 \left( \frac{1}{2}s_x - \frac{\sqrt{3}}{2}s_y,\ \frac{1}{2}s_x + \frac{\sqrt{3}}{2}s_y,\ -s_x \right). \tag{5.7}$$

Then we get

$$\begin{aligned}(\text{the Rashba term}) &= \mathrm{i}\frac{\lambda_R}{a_0} \sum_{\mathbf{k}} \sum_{a=1,2,3} f_a(\mathbf{k}) c_{A\mathbf{k}}^\dagger (\mathbf{s} \times \mathbf{d}_a)_z c_{B\mathbf{k}} + f_a^*(\mathbf{k}) c_{B\mathbf{k}}^\dagger (\mathbf{s} \times (-\mathbf{d}_a))_z c_{A\mathbf{k}} \\ &= \mathrm{i}\lambda_R \sum_{\mathbf{k}} \Bigg[ \left( \frac{1}{2} f_1(\mathbf{k}) + \frac{1}{2} f_2(\mathbf{k}) - f_3(\mathbf{k}) \right) c_{A\mathbf{k}}^\dagger s_x c_{B\mathbf{k}} \\ &\quad + \left( -\frac{\sqrt{3}}{2} f_1(\mathbf{k}) + \frac{\sqrt{3}}{2} f_2(\mathbf{k}) \right) c_{A\mathbf{k}}^\dagger s_y c_{B\mathbf{k}} \\ &\quad - \left( \frac{1}{2} f_1^*(\mathbf{k}) + \frac{1}{2} f_2^*(\mathbf{k}) - f_3^*(\mathbf{k}) \right) c_{B\mathbf{k}}^\dagger s_x c_{A\mathbf{k}} \\ &\quad - \left( -\frac{\sqrt{3}}{2} f_1^*(\mathbf{k}) + \frac{\sqrt{3}}{2} f_2^*(\mathbf{k}) \right) c_{B\mathbf{k}}^\dagger s_y c_{A\mathbf{k}} \Bigg],\end{aligned} \tag{5.8}$$

where $f_a(\mathbf{k}) = e^{\mathrm{i}\mathbf{k}\cdot\mathbf{d}_a}$. To express it in a compact form, we introduce five Dirac matrices: $\Gamma^{1,2,3,4,5} = (\sigma_x \otimes s_0, \sigma_z \otimes s_0, \sigma_y \otimes s_x, \sigma_y \otimes s_y, \sigma_y \otimes s_z)$, where the Pauli matrices $\sigma_i$ and $s_i$ represent the sublattice and spin index, respectively, and we also introduce their ten commutators $\Gamma^{ab} = [\Gamma^a, \Gamma^b]/(2i)$. Therefore, the Bloch Hamiltonian in $\mathbf{k}$ space is represented as

$$\begin{aligned}H_0(\mathbf{k}) &= t_0 F_R(\mathbf{k})\Gamma^1 - t_0 F_I(\mathbf{k})\Gamma^{12} + \lambda_v \Gamma^2 \\ &\quad + \lambda_R F_4(\mathbf{k})\Gamma^{23} - \lambda_R F_2(\mathbf{k})\Gamma^3 - \lambda_R F_3(\mathbf{k})\Gamma^{24} - \lambda_R F_1(\mathbf{k})\Gamma^4,\end{aligned} \tag{5.9}$$

and

$$F_{\mathrm{R}}(\mathbf{k}) = 2\cos x \cos y + \cos 2y, \tag{5.10}$$

$$F_{\mathrm{I}}(\mathbf{k}) = 2\cos x \sin y - \sin 2y, \tag{5.11}$$

$$F_1(\mathbf{k}) = \sqrt{3}\sin x \sin y, \tag{5.12}$$

$$F_2(\mathbf{k}) = \cos x \cos y - \cos 2y, \tag{5.13}$$

$$F_3(\mathbf{k}) = \sqrt{3}\sin x \cos y, \tag{5.14}$$

$$F_4(\mathbf{k}) = \cos x \sin y + \sin 2y, \tag{5.15}$$

with $x = \sqrt{3}k_x a_0/2$ and $y = k_y a_0/2$.

Next, we consider modification of the nearest neighbor hopping term and the nearest neighbor Rashba term by the lattice deformation. A lattice deformation $\mathbf{u}$ modifies the nearest neighbor hopping integral as $t \to t + \delta t_a$ $(a = 1, 2, 3)$ [15]. We define the Hamiltonian for the modulation of the nearest neighbor hopping term as

$$H_{\mathrm{h}}(t) = \sum_{\langle ij \rangle} \delta t_{ij}(t) c_i^{\dagger} c_j. \tag{5.16}$$

We assume that the modulation of the nearest neighbor hopping parameter $\delta t_a$ is proportional to the change of the nearest neighbor bond length, which is given by $\mathbf{u}(t) \cdot (\mathbf{d}_a/a_0)$. Thus, the modulation of the nearest neighbor hopping parameter $\delta t_a(t)$ is given by

$$\delta t_a(t) = -\frac{\delta t_0}{a_0^2}\mathbf{u}(t) \cdot \mathbf{d}_a, \tag{5.17}$$

and each parameter is obtained as

$$\delta t_1(t) = \frac{\delta t_0 u_+}{a_0}\left(\frac{\sqrt{3}}{2}\cos \Omega t + \frac{1}{2}\sin \Omega t\right), \tag{5.18}$$

$$\delta t_2(t) = \frac{\delta t_0 u_+}{a_0}\left(\frac{-\sqrt{3}}{2}\cos \Omega t + \frac{1}{2}\sin \Omega t\right), \tag{5.19}$$

$$\delta t_3(t) = -\frac{\delta t_0 u_+}{a_0}\sin \Omega t. \tag{5.20}$$

Therefore, the Hamiltonian for the modulation of the hopping term in $\mathbf{k}$ space is represented as

$$H_{\mathrm{h}}(\mathbf{k}, t) = -\delta t(F_1(\mathbf{k})\cos \Omega t - F_2(\mathbf{k})\sin \Omega t)\Gamma^1 - \delta t(F_3(\mathbf{k})\cos \Omega t + F_4(\mathbf{k})\sin \Omega t)\Gamma^{12}, \tag{5.21}$$

with $\delta t = \delta t_0 u_+/a_0$. Similarly, we consider the modulation of the nearest neighbor Rashba term. We assume that the Hamiltonian for the modulation of Rashba term

$$H_{\mathrm{R}}(t) = \mathrm{i}\lambda_{\mathrm{R}} \sum_{\langle ij \rangle} c_i^\dagger \left( \mathbf{s} \times \frac{\mathbf{d}'_{ij}}{|d'_{ij}|} \right)_z c_j, \tag{5.22}$$

and this Hamiltonian can be expanded in a Taylor series in powers of the displacement vector $\mathbf{u}$

$$\begin{aligned} H_{\mathrm{R}}(t) = & \frac{\mathrm{i}\lambda_R}{a_0} \sum_{\langle ij \rangle} c_i^\dagger (\mathbf{s} \times \mathbf{d}_{ij})_z c_j + \sum_{\langle ij \rangle} \frac{-\mathrm{i}\lambda_{\mathrm{R}}}{a_0^3} (\mathbf{d}_{ij} \cdot \mathbf{u}(t)) c_i^\dagger (\mathbf{s} \times \mathbf{d}_{ij}) c_j \\ & + \frac{\mathrm{i}\lambda_{\mathrm{R}}}{a_0} \sum_{\langle ij \rangle} c_i^\dagger (\mathbf{s} \times \mathbf{u}(t))_z c_j + O(\mathbf{u}^2). \end{aligned} \tag{5.23}$$

The first term is the Rashba term without the periodic microscopic local rotation and the second and third terms are the modulated Rashba terms by the displacement vector $\mathbf{u}$. The modulated Rashba term $H'_{\mathrm{R}}(t)$ for the wave vector $\mathbf{k}$ is rewritten as

$$\begin{aligned} H'_{\mathrm{R}}(\mathbf{k}, t) = & \delta\lambda_{\mathrm{R}} [(-F_3(\mathbf{k}) \cos \Omega t - (2F_{\mathrm{I}}(\mathbf{k}) + F_5(\mathbf{k})) \sin \Omega t)\Gamma^{23} \\ & + (-F_1(\mathbf{k}) \cos \Omega t + (2F_{\mathrm{R}}(\mathbf{k}) + F_6(\mathbf{k})) \sin \Omega t)\Gamma^3 \\ & + ((2F_{\mathrm{I}}(\mathbf{k}) + 3 \cos x \sin y) \cos \Omega t + F_3(\mathbf{k}) \sin \Omega t)\Gamma^{24} \\ & + (-(2F_{\mathrm{R}}(\mathbf{k}) + 3 \cos x \cos y) \cos \Omega t + F_1(\mathbf{k}) \sin \Omega t)\Gamma^4], \end{aligned} \tag{5.24}$$

and

$$F_5(\mathbf{k}) = \cos x \sin y - 2 \sin 2y, \tag{5.25}$$

$$F_6(\mathbf{k}) = \cos x \cos y + 2 \cos 2y, \tag{5.26}$$

with $\delta\lambda_{\mathrm{R}} = \lambda_{\mathrm{R}} u_+/(2a_0)$. Therefore, the Hamiltonian for periodic microscopic local rotation in $\mathbf{k}$ space is represented as

$$H(\mathbf{k}, t) = H_0(\mathbf{k}) + H_{\mathrm{h}}(\mathbf{k}, t) + H'_{\mathrm{R}}(\mathbf{k}, t). \tag{5.27}$$

## 5.3 Adiabatic Series Expansion

We need to solve the time-dependent Schrödinger equation with the time-periodic Hamiltonian. In general, the motions of atoms for phonons are much slower than those of electrons because the masses of atoms are much larger than those of electrons. Therefore, in our toy model, we assume that under the rotational motions of atoms the electronic states of our toy model change adiabatically with respect to time, and the electron in one state does not transit to other states under time evolution. In this section, we explain the method of the adiabatic series expansion proposed by Berry [16, 17].

First, we introduce a time-periodic Hamiltonian and its time-dependent Schrödinger equation

$$H(t) = H_0 + H_1(t), \tag{5.28}$$

$$i\hbar\frac{d}{dt}\psi(t) = H(t)\psi(t), \tag{5.29}$$

where $\psi(t)$ is a wave function. This Hamiltonian satisfies $H(t+T) = H(t)$ with a period $T = 2\pi/\Omega$. Here, we introduce the rescaled time $\tau = \Omega t$, and rewrite the functions of time as $\psi(t) = \psi(\tau),\ H(t) = H(\tau),\ H(\tau + 2\pi) = H(\tau)$. Therefore, the time-dependent Schrödinger equation is rewritten as

$$i\hbar\Omega\frac{d}{d\tau}\psi(\tau) = H(\tau)\psi(\tau). \tag{5.30}$$

Next, we introduce the time-evolution operator $U$, which is unitary operator. In terms of the time-evolution operator $U(\tau)$, the wave function is written as

$$\psi(\tau) = U(\tau)\psi(0). \tag{5.31}$$

By substituting this into Eq. (5.30), we get

$$i\hbar\Omega\partial_\tau U(\tau)\psi(0) = H(\tau)U(\tau)\psi(0), \tag{5.32}$$

$$[U^\dagger(\tau)H(\tau)U(\tau) - i\hbar\Omega U^\dagger(\tau)\partial_\tau U(\tau)]\psi(0) = 0. \tag{5.33}$$

Here, we introduce the instantaneous eigenvalues $\epsilon_n(\tau)$ and the instantaneous eigenstates $|\epsilon_n(\tau)\rangle$ of the Hamiltonian $H(\tau)$,

$$H(\tau)\,|\epsilon_n(\tau)\rangle = \epsilon_n(\tau)\,|\epsilon_n(\tau)\rangle\,, \tag{5.34}$$

where $n$ is the band index. In the following, we will calculate $U(\tau)$ in the form

$$U(\tau) = R_0(\tau)R_1(\tau)R_2(\tau)\cdots, \tag{5.35}$$

where $R_i(\tau)\ (i = 1, 2, 3, \ldots)$ is an unitary operator. As a zeroth-order approximation of $U(\tau)$. We define the unitary operator $R_0(\tau)$ as

$$R_0(\tau) = \sum_n |\epsilon_n(\tau)\rangle\,\langle\epsilon_n(0)|\,. \tag{5.36}$$

Time derivative of $R_0(\tau)$ is written as

$$\partial_\tau R_0(\tau) = \sum_n |\partial_\tau \epsilon_n(\tau)\rangle \langle \epsilon_n(0)| = \sum_n |\partial_\tau \epsilon_n(\tau)\rangle \langle \epsilon_n(\tau)| R_0(\tau). \tag{5.37}$$

Then we rewrite the Eq. (5.33),

$$\begin{aligned} & U^\dagger[H(\tau) - \mathrm{i}\hbar\Omega\partial_\tau]U \\ &= \cdots R_2^\dagger R_1^\dagger R_0^\dagger[H(\tau) - \mathrm{i}\hbar\Omega\partial_\tau]R_0 R_1 R_2 \cdots \\ &= \cdots R_2^\dagger R_1^\dagger[R_0^\dagger H(\tau) R_0 - \mathrm{i}\hbar\Omega R_0^\dagger \partial_\tau R_0 - \mathrm{i}\hbar\Omega\partial_\tau]R_1 R_2 \cdots \\ &= \cdots R_2^\dagger R_1^\dagger[(\tilde{H}^{(0)}(\tau) + \Omega\tilde{V}^{(0)}(\tau))R_1 - \mathrm{i}\hbar\Omega\partial_\tau R_1 - \mathrm{i}\hbar\Omega R_1\partial_\tau]R_2 \cdots, \end{aligned} \tag{5.38}$$

and

$$\begin{aligned} \tilde{H}^{(0)}(\tau) &= R_0^\dagger H(\tau) R_0 \\ &= \sum_{n,m} |\epsilon_{km}(0)\rangle \langle \epsilon_{km}(\tau)| H(\tau) |\epsilon_{kn}(\tau)\rangle \langle \epsilon_{kn}(0)| \\ &= \sum_n \epsilon_{kn}(\tau) |\epsilon_{kn}(0)\rangle \langle \epsilon_{kn}(0)|, \end{aligned} \tag{5.39}$$

$$\tilde{V}^{(0)}(\tau) = -\mathrm{i}\hbar R_0^\dagger \partial_\tau R_0. \tag{5.40}$$

Note that time-dependence of $\tilde{H}^{(0)}(\tau)$ is contained only in the eigenvalues $\epsilon_n(\tau)$.

In the first-order approximation of the adiabatic series expansion, we consider that $\Omega\tilde{V}^{(0)}(\tau)$ is much smaller than $\tilde{H}^{(0)}(\tau)$ $[\Omega\tilde{V}^{(0)} \ll \tilde{H}^{(0)}]$, and the $\Omega\tilde{V}^{(0)}(\tau)$ term is dropped. Here, we put the unitary operators $R_2 = R_3 = \cdots = I$, where $I$ is an identity operator, and we calculate $R_1$. Then, from Eq. (5.38), the solution of $\tilde{H}^{(0)}(\tau)R_1(\tau) - \mathrm{i}\hbar\Omega\partial_\tau R_1(\tau) = 0$ is given by

$$R_1(\tau) = R_{1,d}(\tau) = \exp\left(\frac{1}{\mathrm{i}\hbar\Omega}\int_0^t ds\,\tilde{H}^{(0)}(s)\right). \tag{5.41}$$

In the first-order approximation, the state at $\tau$ is represented as

$$\psi(\tau) = U(\tau)\psi(0) = R_0(\tau)R_{1,d}(\tau)\psi(0). \tag{5.42}$$

Next, we consider the second-order approximation of the adiabatic series expansion. Here, we put $\tilde{H}_1(\tau) = \tilde{H}^{(0)}(\tau) + \Omega\tilde{V}^{(0)}$, and we set the instantaneous eigenvalues $E_n(\tau)$ and the instantaneous eigenstate $|\phi_n(\tau)\rangle$ of $\tilde{H}_1(\tau)$,

$$\tilde{H}_1(\tau)|\phi_n(\tau)\rangle = E_n(\tau)|\phi_n(\tau)\rangle. \tag{5.43}$$

We perform the perturbation expansion for $\tilde{H}_1(\tau)$, and $E_n(\tau)$ and $|\phi_n(\tau)\rangle$ is represented to the first order of $\Omega$ as

$$E_n(\tau) = \epsilon_n(\tau) + \Omega \langle \epsilon_n(0)| \tilde{V}^{(0)}(\tau) |\epsilon_n(0)\rangle , \tag{5.44}$$

$$|\phi_n(\tau)\rangle = |\epsilon_n(0)\rangle + \Omega \sum_{m \neq n} \frac{\langle \epsilon_m(0)| \tilde{V}^{(0)}(\tau) |\epsilon_n(0)\rangle}{\epsilon_n(\tau) - \epsilon_m(\tau)} |\epsilon_m(0)\rangle . \tag{5.45}$$

Note that we assume that the instantaneous eigenvalues $\epsilon_n(\tau)$ for the instantaneous eigenstates $|\epsilon_n(\tau)\rangle$ have no degeneracy. We define the unitary operator $R_1(\tau)$ as

$$R_1(\tau) = \sum_n |\phi_n(\tau)\rangle \langle \phi_n(0)| , \tag{5.46}$$

and its time derivative is given by

$$\partial_\tau R_1(\tau) = \sum_n |\partial_\tau \phi_n(\tau)\rangle \langle \phi_n(0)| = \sum_n |\partial_\tau \phi_n(\tau)\rangle \langle \phi_n(\tau)| R_1(\tau). \tag{5.47}$$

From Eq. (5.38), one has

$$\begin{aligned} & U^\dagger[H(\tau) - i\hbar\Omega\partial_\tau]U \\ & = \cdots R_2^\dagger R_1^\dagger[(\tilde{H}^{(0)}(\tau) + \Omega\tilde{V}^{(0)}(\tau))R_1 - \mathrm{i}\hbar\Omega\partial_\tau R_1 - \mathrm{i}\hbar\Omega R_1\partial_\tau]R_2 \cdots \\ & = \cdots R_2^\dagger[R_1^\dagger \tilde{H}_1(\tau)R_1 - \mathrm{i}\hbar\Omega R_1^\dagger \partial_\tau R_1 - \mathrm{i}\hbar\Omega\partial_\tau]R_2 \cdots \\ & = \cdots R_2^\dagger[(\tilde{H}^{(1)}(\tau) + \Omega\tilde{V}^{(1)}(\tau))R_2 - \mathrm{i}\hbar\Omega\partial_\tau R_2 - \mathrm{i}\hbar\Omega R_2\partial_\tau]R_3 \cdots , \end{aligned} \tag{5.48}$$

and

$$\begin{aligned} \tilde{H}^{(1)}(\tau) &= R_1^\dagger \tilde{H}_1(\tau) R_1 \\ &= \sum_{n,m} |\phi_{km}(0)\rangle \langle \phi_{km}(\tau)| \tilde{H}_1(\tau) |\phi_{kn}(\tau)\rangle \langle \phi_{kn}(0)| \\ &= \sum_n E_{kn}(\tau) |\phi_{kn}(0)\rangle \langle \phi_{kn}(0)| , \end{aligned} \tag{5.49}$$

$$\tilde{V}^{(1)}(\tau) = -\mathrm{i}\hbar R_1^\dagger \partial_\tau R_1. \tag{5.50}$$

Note that time-dependence of $\tilde{H}^{(1)}(\tau)$ is contained only in the eigenvalues $E_n(\tau)$

In the second-order approximation of the adiabatic series expansion, we assume that $\Omega\tilde{V}^{(1)}(\tau)$ is of the order $O(\Omega^2)$ and is sufficiently small. Therefore, the $\Omega\tilde{V}^{(1)}(\tau)$ term is dropped. We put the unitary operators $R_3 = R_4 = \cdots = I$ and calculate $R_2$. Then, from Eq. (5.48), the solution of $\tilde{H}^{(1)}(\tau)R_2(\tau) - \mathrm{i}\hbar\Omega\partial_\tau R_2(\tau) = 0$ is given by

$$R_2(\tau) \equiv R_{2,d} = \exp\left(\frac{1}{\mathrm{i}\hbar\Omega}\int_0^\tau ds \tilde{H}^{(1)}(s)\right). \tag{5.51}$$

In the second-order approximation, the state at time $\tau$ is represented as

$$\psi(\tau) = U(\tau)\psi(0) = R_0(\tau)R_1(\tau)R_{2,d}(\tau)\psi(0). \tag{5.52}$$

Similarly, we can iterate this expansion, and then the state at $\tau$ is obtained as

$$\psi(\tau) = U(\tau)\psi(0) = R_0(\tau)R_1(\tau)\cdots R_{n,d}(\tau)\psi(0). \tag{5.53}$$

## 5.4 Expectation Values in the Adiabatic Approximation

Since we are interested in the effect of a mechanical rotation, we expand the expectation value of an arbitrary operator up to the first order of the angular velocity $\Omega$. We will write down the expectation value of an arbitrary operator $\hat{O}$ in the second-order approximation in the adiabatic series expansion. The expectation value $\langle O(\tau)_{nm}\rangle$ is given by

$$\begin{aligned}\langle O(\tau)_{nm}\rangle &= \langle\psi_n(\tau)|\,\hat{O}\,|\psi_m(\tau)\rangle \\ &= \langle\psi_n(0)|\,R^\dagger_{2,d}(\tau)R^\dagger_1(\tau)R^\dagger_0(\tau)\hat{O}R_0(\tau)R_1(\tau)R_{2,d}(\tau)\,|\psi_m(0)\rangle\,,\end{aligned} \tag{5.54}$$

where $|\psi_n(0)\rangle$ is the initial state. First, the unitary operator $R_{2,d}(\tau)$ is rewritten as

$$\begin{aligned}R_{2,d}(\tau) &= \exp\left(\frac{-\mathrm{i}}{\hbar\Omega}\int_0^\tau ds\tilde{H}^{(1)}(s)\right) \\ &= \exp\left(\frac{-\mathrm{i}}{\hbar\Omega}\int_0^\tau ds\sum_n E_n(s)\,|\phi_n(0)\rangle\,\langle\phi_n(0)|\right) \\ &= \exp\left(\frac{-\mathrm{i}}{\hbar\Omega}\sum_n \overline{E}_n(s)\,|\phi_n(0)\rangle\,\langle\phi_n(0)|\right) \\ &= \sum_n \exp\left(\frac{-\mathrm{i}}{\hbar\Omega}\overline{E}_n(s)\right)|\phi_n(0)\rangle\,\langle\phi_n(0)|\,.\end{aligned} \tag{5.55}$$

In the third equality, we introduced

$$\overline{E}_n(\tau) = \int_0^\tau dsE_n(s). \tag{5.56}$$

In the fourth equality, we used the following relation

$$\exp\left(\sum_n A_n |\phi_n(0)\rangle \langle\phi_n(0)|\right) = 1 + \sum_n A_n |\phi_n(0)\rangle \langle\phi_n(0)|$$
$$+ \frac{1}{2}\sum_n\sum_m A_n A_m (|\phi_n(0)\rangle \langle\phi_n(0)|)(|\phi_m(0)\rangle \langle\phi_m(0)|)$$
$$+ \cdots$$
$$= \sum_{m=0}^{\infty} \frac{1}{m!}\sum_n (A_n)^m |\phi_n(0)\rangle \langle\phi_n(0)|$$
$$= \sum_n \exp(A_n) |\phi_n(0)\rangle \langle\phi_n(0)|, \tag{5.57}$$

from the orthogonality relation $\langle\phi_n(\tau)|\phi_m(\tau) = \delta_{nm}\rangle$. Then, we can rewrite the $R_1(\tau)R_{2,d}(\tau)$, $R^\dagger_{2,d}(\tau)R^\dagger_1(\tau)$ and $R^\dagger_0(\tau)\hat{O}R_0(\tau)$ as

$$R_1(\tau)R_{2,d}(\tau) = \sum_{n,m} |\phi_n(\tau)\rangle \langle\phi_n(0)| \exp\left(\frac{-\mathrm{i}\overline{E}_m(\tau)}{\hbar\Omega}\right) |\phi_m(0)\rangle \langle\phi_m(0)|$$
$$= \sum_n \exp\left(\frac{-\mathrm{i}\overline{E}_n(\tau)}{\hbar\Omega}\right) |\phi_n(\tau)\rangle \langle\phi_n(0)|, \tag{5.58}$$
$$R^\dagger_{2,d}(\tau)R^\dagger_1(\tau) = \sum_n \exp\left(\frac{\mathrm{i}\overline{E}_n(\tau)}{\hbar\Omega}\right) |\phi_n(0)\rangle \langle\phi_n(\tau)|, \tag{5.59}$$
$$R^\dagger_0(\tau)\hat{O}R_0(\tau) = \sum_{nm} |\epsilon_n(0)\rangle \langle\epsilon_n(\tau)| \hat{O} |\epsilon_m(\tau)\rangle \langle\epsilon_m(0)|$$
$$= \sum_{nm} O^{nm}(\tau) |\epsilon_n(0)\rangle \langle\epsilon_m(0)|. \tag{5.60}$$

Hence, we obtain

$$R^\dagger_{2,d}(\tau)R^\dagger_1(\tau)R^\dagger_0(\tau)\hat{O}R_0(\tau)R_1(\tau)R_{2,d}(\tau)$$
$$= \sum_{n,m} \exp\left(\mathrm{i}\frac{\overline{E}_n(\tau) - \overline{E}_m(\tau)}{\hbar\Omega}\right) \langle\phi_n(\tau)| R^\dagger_0(\tau)\hat{O}R_0(\tau) |\phi_m(\tau)\rangle |\phi_n(0)\rangle \langle\phi_m(0)|. \tag{5.61}$$

Here, from Eq. (5.45), $\langle\phi_n(\tau)| R^\dagger_0(\tau)\hat{O}R_0(\tau) |\phi_m(\tau)\rangle$ and $|\phi_n(0)\rangle \langle\phi_m(0)|$ can be written down as

$$
\begin{aligned}
\langle\phi_n(\tau)|\, R_0^{\dagger}(\tau)\hat{O}R_0(\tau)\,|\phi_m(\tau)\rangle &= \langle\epsilon_n(0)|\, R_0^{\dagger}(\tau)\hat{O}R_0(\tau)\,|\epsilon_m(0)\rangle \\
&+ \Omega\sum_{l\neq m}\frac{\langle\epsilon_l(0)|\,\tilde{V}^{(0)}(\tau)\,|\epsilon_m(0)\rangle}{\epsilon_m(\tau)-\epsilon_l(\tau)}\langle\epsilon_n(0)|\, R_0^{\dagger}(\tau)\hat{O}R_0(\tau)\,|\epsilon_l(0)\rangle \\
&+ \Omega\sum_{k\neq n}\frac{\langle\epsilon_n(0)|\,\tilde{V}^{(0)}(\tau)\,|\epsilon_k(0)\rangle}{\epsilon_n(\tau)-\epsilon_k(\tau)}\langle\epsilon_k(0)|\, R_0^{\dagger}(\tau)\hat{O}R_0(\tau)\,|\epsilon_m(0)\rangle \\
&+ O(\Omega^2) \\
&= O^{nm}(\tau) + \Omega\sum_{l\neq m}\frac{\langle\epsilon_l(0)|\,\tilde{V}^{(0)}(\tau)\,|\epsilon_m(0)\rangle}{\epsilon_m(\tau)-\epsilon_l(\tau)}O^{nl}(\tau) \\
&+ \Omega\sum_{k\neq n}\frac{\langle\epsilon_n(0)|\,\tilde{V}^{(0)}(\tau)\,|\epsilon_k(0)\rangle}{\epsilon_n(\tau)-\epsilon_k(\tau)}O^{km}(\tau) + O(\Omega^2),
\end{aligned}
\tag{5.62}
$$

$$
\begin{aligned}
|\phi_n(0)\rangle\langle\phi_m(0)| &= |\epsilon_n(0)\rangle\langle\epsilon_m(0)| \\
&+ \Omega\sum_{\alpha\neq m}\frac{\langle\epsilon_m(0)|\,\tilde{V}^{(0)}(\tau)\,|\epsilon_\alpha(0)\rangle}{\epsilon_m(\tau)-\epsilon_\alpha(\tau)}|\epsilon_n(0)\rangle\langle\epsilon_\alpha(0)| \\
&+ \Omega\sum_{\beta\neq n}\frac{\langle\epsilon_\beta(0)|\,\tilde{V}^{(0)}(\tau)\,|\epsilon_n(0)\rangle}{\epsilon_n(\tau)-\epsilon_\beta(\tau)}|\epsilon_\beta(0)\rangle\langle\epsilon_m(0)| + O(\Omega^2).
\end{aligned}
\tag{5.63}
$$

Here, we drop the terms of the order $O(\Omega^2)$, and the expectation value of the operator $\hat{O}$ is given by

$$
\begin{aligned}
&\langle\epsilon_i(0)|\, R_{2,d}^{\dagger}(\tau)R_1^{\dagger}(\tau)R_0^{\dagger}(\tau)\hat{O}R_0(\tau)R_1(\tau)R_{2,d}(\tau)\,|\epsilon_j(0)\rangle \\
&= \sum_{n,m}\exp\left(\mathrm{i}\frac{\overline{E}_n(\tau)-\overline{E}_m(\tau)}{\hbar\Omega}\right)\langle\phi_n(\tau)|\, R_0^{\dagger}(\tau)\hat{O}R_0(\tau)\,|\phi_m(\tau)\rangle\,\langle|\epsilon_i(0)|\phi_n(0)\rangle\langle|\phi_m(0)|\epsilon_j(0)\rangle \\
&= \exp\left(\mathrm{i}\frac{\overline{E}_i(\tau)-\overline{E}_j(\tau)}{\hbar\Omega}\right)\left[O^{ij}(\tau) + \Omega\sum_{l\neq j}\frac{\langle\epsilon_l(0)|\,\tilde{V}^{(0)}(\tau)\,|\epsilon_j(0)\rangle}{\epsilon_j(\tau)-\epsilon_l(\tau)}O^{il}(\tau)\right. \\
&\qquad\left. + \Omega\sum_{k\neq i}\frac{\langle\epsilon_i(0)|\,\tilde{V}^{(0)}(\tau)\,|\epsilon_k(0)\rangle}{\epsilon_i(\tau)-\epsilon_k(\tau)}O^{kj}(\tau)\right] \\
&+ \Omega\sum_{m\neq j}\exp\left(\mathrm{i}\frac{\overline{E}_i(\tau)-\overline{E}_m(\tau)}{\hbar\Omega}\right)\frac{\langle\epsilon_m(0)|\,\tilde{V}^{(0)}(\tau)\,|\epsilon_j(0)\rangle}{\epsilon_m(\tau)-\epsilon_j(\tau)}O^{im}(\tau) \\
&+ \Omega\sum_{n\neq i}\exp\left(\mathrm{i}\frac{\overline{E}_n(\tau)-\overline{E}_j(\tau)}{\hbar\Omega}\right)\frac{\langle\epsilon_i(0)|\,\tilde{V}^{(0)}(\tau)\,|\epsilon_n(0)\rangle}{\epsilon_n(\tau)-\epsilon_i(\tau)}O^{nj}(\tau).
\end{aligned}
\tag{5.64}
$$

In this section, we use the adiabatic series expansion proposed by Berry [16, 17]. Recently, another method to calculate the expectation values in the adiabatic approximation has been proposed, and the form of the time-averaged expectation values by using the another method matches our results [18].

## 5.5 Expectation Values of Spin Operator in Our Model

In this section, we discuss expectation values of the spin operators in the adiabatic approximation for our model. The spin operator is expressed as

$$\hat{S}_i = \sigma_0 \otimes s_i \quad (i = x, y, z) \tag{5.65}$$

with the identity matrix $\sigma_0$. First, we calculate the spin expectation values without the microscopic local rotation. From the Hamiltonian in **k** space, Eq. (5.9), the electronic energy bands on a high-symmetry line is shown in Fig. 5.2a. We assume that the Fermi energy $E_F$ is zero, and the temperature is zero. Our toy model has a gap since the staggered sublattice potential term breaks the inversion symmetry, and the electron spin is split by the Rashba spin-orbit term. The spin structure below the Fermi energy $E_F$ in the Brillouin zone is shown in Fig. 5.2c, d. Since the time-reversal symmetry is conserved, the spin expectation values are odd functions of wave vector **k**, and their sums over the Brillouin zone vanish. Hence, in our toy model without the microscopic local rotation, the spin expectation values below the Fermi energy $E_F$ are zero, and this model is not magnetized.

Next, we show the spin expectation values with the periodic microscopic local rotation of atoms. Here, we replace the time $t$ to the rescaled time $\tau = \Omega t$ in Eq. (5.27). By using the second order adiabatic approximation, the expectation value of operator $\hat{O}$ is represented as

$$\begin{aligned}
&\langle \hat{O}_n(\mathbf{k}, \tau) \rangle \\
&= \langle \epsilon_n(\mathbf{k}, \tau) | \, \hat{O} \, | \epsilon_n(\mathbf{k}, \tau) \rangle \\
&= \langle \epsilon_n(\mathbf{k}, 0) | \, R_{2,d}(\mathbf{k}, \tau)^\dagger R_1(\mathbf{k}, \tau)^\dagger R_0(\mathbf{k}, \tau)^\dagger \hat{O} R_0(\mathbf{k}, \tau) R_1(\mathbf{k}, \tau) R_{2,d}(\mathbf{k}, \tau) \, | \epsilon_n(\mathbf{k}, 0) \rangle ,
\end{aligned} \tag{5.66}$$

and the unitary operators $R_0(\mathbf{k}, \tau)$, $R_1(\mathbf{k}, \tau)$ and $R_{2,d}(\mathbf{k}, \tau)$ are obtained from Eqs. (5.36), (5.46), and (5.55) as

$$R_0(\mathbf{k}, \tau) = \sum_{n=1}^{4} |\epsilon_n(\mathbf{k}, \tau)\rangle \langle \epsilon_n(\mathbf{k}, 0)| , \tag{5.67}$$

$$R_1(\mathbf{k}, \tau) = \sum_{n=1}^{4} |\phi_n(\mathbf{k}, \tau)\rangle \langle \phi_n(\mathbf{k}, 0)| , \tag{5.68}$$

$$R_{2,d}(\mathbf{k}, \tau) = \sum_{n=1}^{4} \exp\left( \frac{-i}{\hbar\Omega} \int_0^\tau E_n(\mathbf{k}, \tau) \right) |\phi_n(\mathbf{k}, 0)\rangle \langle \phi_n(\mathbf{k}, 0)| , \tag{5.69}$$

and the instantaneous eigenvalues $E_n$ and the instantaneous eigenstates $|\phi_n\rangle$ are given from Eqs. (5.44) and (5.45)

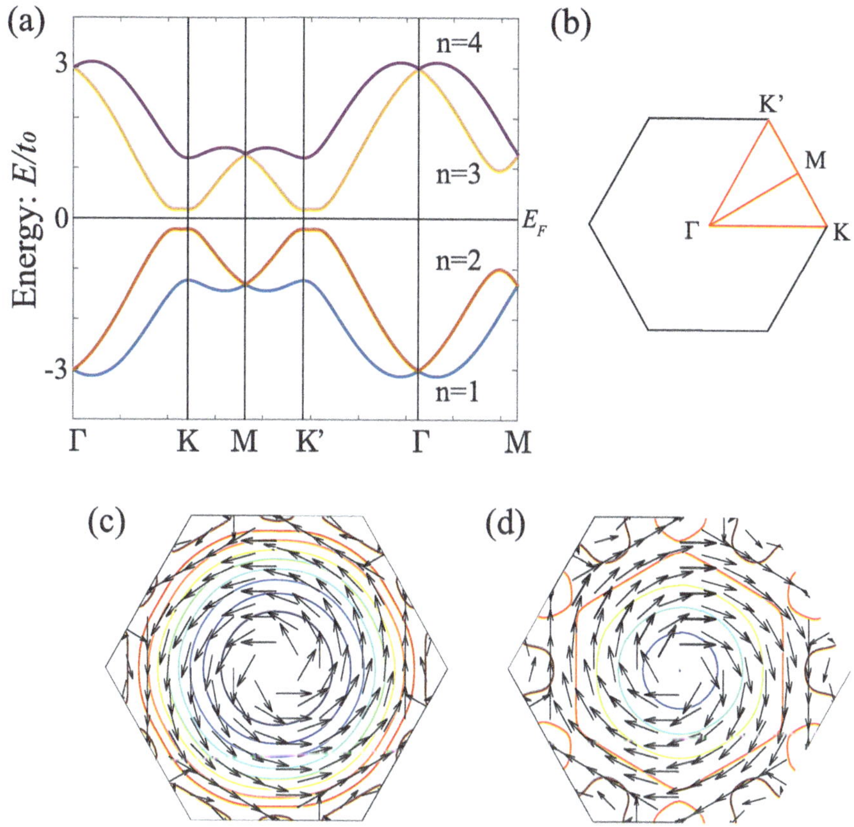


**Fig. 5.2** Electronic energy bands in the absence of the microscopic local rotation and the electron spin structure below the Fermi energy. **a** Energy bands without the microscopic local rotation on the high symmetry line. **b** First Brillouin zone in our model. **c** Spin structure of the lowest band. **d** Spin structure of the second lowest band. In **c** and **d**, the black arrows show the spin expectation vector $\mathbf{S}(\mathbf{k}) = (S_x(\mathbf{k}), S_y(\mathbf{k}))$, and the color lines show the equal energy lines. We set the parameters as $\lambda_v = 0.2t_0$, $\lambda_R = 0.4t_0$

$$E_n(\mathbf{k}, \tau) = \epsilon_n(\mathbf{k}, \tau) + \hbar\Omega\langle\epsilon_n(\mathbf{k}, \tau)|(-\mathrm{i})\partial_\tau\epsilon_n(\mathbf{k}, \tau)\rangle, \tag{5.70}$$

$$|\phi_n(\mathbf{k}, \tau)\rangle = |\epsilon_n(\mathbf{k}, 0)\rangle + \hbar\Omega\sum_{m\neq n}\frac{\langle\epsilon_m(\mathbf{k}, \tau)|(-\mathrm{i})\partial_\tau\epsilon_n(\mathbf{k}, \tau)\rangle}{\epsilon_n(\mathbf{k}, \tau) - \epsilon_m(\mathbf{k}, \tau)}|\epsilon(\mathbf{k}, 0)\rangle\,. \tag{5.71}$$

Here, we introduce an expression for $\langle\epsilon_n(\mathbf{k}, \tau)|\partial_\tau\epsilon_m(\mathbf{k}, \tau)\rangle$. By differentiating Eq. (5.34) with respect to the rescaled time $\tau$, one has

$$[\partial_\tau H(\mathbf{k}, \tau) - \partial_\tau\epsilon_n(\mathbf{k}, \tau)]\,|\epsilon_n(\mathbf{k}, \tau)\rangle = [\epsilon_n(\mathbf{k}, \tau) - H(\mathbf{k}, \tau)]\,|\partial_\tau\epsilon_n(\mathbf{k}, \tau)\rangle\,. \tag{5.72}$$

We multiply $\langle \epsilon_m(\mathbf{k}, \tau)|$ $(m \neq n)$ from the left, and we use the orthogonality relation $\langle \epsilon_n(\mathbf{k}, \tau)|\epsilon_m(\mathbf{k}, \tau)\rangle = \delta_{nm}$. Then, we get

$$\langle \epsilon_m(\mathbf{k}, \tau)|\partial_\tau \epsilon_n(\mathbf{k}, \tau)\rangle = \frac{\langle \epsilon_m(\mathbf{k}, \tau)|\, \partial_\tau H(\mathbf{k}, \tau)\, |\epsilon_n(\mathbf{k}, \tau)\rangle}{\epsilon_n(\mathbf{k}, \tau) - \epsilon_m(\mathbf{k}, \tau)} \quad (n \neq m). \tag{5.73}$$

We also show an expression for $\langle \epsilon_n(\mathbf{k}, \tau)|\partial_\tau \epsilon_n(\mathbf{k}, \tau)\rangle$. Differentiation of the orthogonality relation with respect to the rescaled time $\tau$ gives

$$\langle \partial_\tau \epsilon_n(\mathbf{k}, \tau)|\epsilon_n(\mathbf{k}, \tau)\rangle + \langle \epsilon_n(\mathbf{k}, \tau)|\partial_\tau \epsilon_n(\mathbf{k}, \tau)\rangle = 0. \tag{5.74}$$

Since these two terms are complex conjugates of each other, both are imaginary, we can write $\langle \epsilon_n(\mathbf{k}, \tau)|\partial_\tau \epsilon_n(\mathbf{k}, \tau)\rangle = i\gamma_n(\mathbf{k}, \tau)$, where $\gamma_n(\mathbf{k}, \tau)$ is real. $\gamma_n(\mathbf{k}, \tau)$ is a geometrical phase, and this is called a Berry phase. We can rewrite Eqs. (5.70) and (5.71) as

$$E_n(\mathbf{k}, \tau) = \epsilon_n(\mathbf{k}, \tau) + \hbar\Omega\gamma_n(\mathbf{k}, \tau), \tag{5.75}$$

$$|\phi_n(\mathbf{k}, \tau)\rangle = |\epsilon_n(\mathbf{k}, 0)\rangle + \hbar\Omega \sum_{m\neq n} \frac{(-\mathrm{i})\langle \epsilon_m(\mathbf{k}, \tau)|\partial_\tau H(\mathbf{k}, \tau)|\epsilon_n(\mathbf{k}, \tau)\rangle}{(\epsilon_n(\mathbf{k}, \tau) - \epsilon_m(\mathbf{k}, \tau))^2} |\epsilon_m(\mathbf{k}, 0)\rangle . \tag{5.76}$$

From Eqs. (5.64), (5.66) can be written as

$$\begin{aligned}
&\langle \hat{O}_n(\mathbf{k}, \tau)\rangle \\
&= O^{nn}(\mathbf{k}, \tau) \\
&+ \hbar\Omega \sum_{m\neq n} \frac{(-\mathrm{i})\, \langle \epsilon_m(\mathbf{k}, \tau)|\, \partial_\tau H(\mathbf{k}, \tau)\, |\epsilon_n(\mathbf{k}, \tau)\rangle\, O^{nm}(\mathbf{k}, \tau)}{(\epsilon_n(\mathbf{k}, \tau) - \epsilon_m(\mathbf{k}, \tau))^2} (1 - e^{\frac{\mathrm{i}}{\hbar\Omega}[\overline{E}_n(\mathbf{k},\tau) - \overline{E}_m(\mathbf{k},\tau)]}) \\
&+ \hbar\Omega \sum_{m\neq n} \frac{\mathrm{i}\, \langle \epsilon_n(\mathbf{k}, \tau)|\, \partial_\tau H(\mathbf{k}, \tau)\, |\epsilon_m(\mathbf{k}, \tau)\rangle\, O^{mn}(\mathbf{k}, \tau)}{(\epsilon_n(\mathbf{k}, \tau) - \epsilon_m(\mathbf{k}, \tau))^2} (1 - e^{\frac{-\mathrm{i}}{\hbar\Omega}[\overline{E}_n(\mathbf{k},\tau) - \overline{E}_m(\mathbf{k},\tau)]}),
\end{aligned} \tag{5.77}$$

with $\overline{E}_n(\mathbf{k}, \tau) = \int_0^\tau ds E_n(\mathbf{k}, s)$. The phase terms in the second and the third terms are sums of the dynamical phase and the Berry phase. Since we have assumed $\epsilon_n(\mathbf{k}, \tau)/\hbar\Omega \gg 1$, these phase terms rapidly oscillate, and these terms can be neglected after time averaging. The energy expectation value is given by

$$\langle \epsilon_n(\mathbf{k}, \tau)\rangle = \langle \epsilon_n(\mathbf{k}, \tau)|\, H(\mathbf{k}, \tau)\, |\epsilon_n(\mathbf{k}, \tau)\rangle = \epsilon_n(\mathbf{k}, \tau). \tag{5.78}$$

Since the off-diagonal term $\langle \epsilon_n(\mathbf{k}\tau)|\, H(\mathbf{k}, \tau)\, |\epsilon_m(\mathbf{k}, \tau)\rangle$ is zero, the energy expectation value is equal to the instantaneous eigenvalues $\epsilon_n(\mathbf{k}, \tau)$. Therefore, each energy band changes slightly with the respect to the rescaled time $\tau$ and the energy gap remains open. Next, the spin expectation values are represented as

$$\begin{aligned}\langle S_{i,n}(\mathbf{k},\tau)\rangle &= S_i^{nn}(\mathbf{k},\tau)\\ &+ \hbar\Omega \sum_{m\neq n} \frac{(-\mathrm{i})\,\langle \epsilon_m(\mathbf{k},\tau)|\,\partial_\tau H(\mathbf{k},\tau)\,|\epsilon_n(\mathbf{k},\tau)\rangle\, S_i^{nm}(\mathbf{k},\tau)}{(\epsilon_n(\mathbf{k},\tau)-\epsilon_m(\mathbf{k},\tau))^2}\\ &+ \hbar\Omega \sum_{m\neq n} \frac{\mathrm{i}\,\langle \epsilon_n(\mathbf{k},\tau)|\,\partial_\tau H(\mathbf{k},\tau)\,|\epsilon_m(\mathbf{k},\tau)\rangle\, S_i^{mn}(\mathbf{k},\tau)}{(\epsilon_n(\mathbf{k},\tau)-\epsilon_m(\mathbf{k},\tau))^2}.\end{aligned} \tag{5.79}$$

Since the energy expectation values change slightly with the respect to the rescaled time $\tau$, the energy bands below the Fermi energy $E_F$ remains almost the same with respect to the rescaled time $\tau$. We show the sum of spin expectation values over the first Brillouin zone for the each energy band below the Fermi energy $E_F$ in Fig. 5.3. In Fig. 5.3, the first term in Eq. (5.79) is zero for all the components, and the second and the third terms are nonzero. The spin expectation values of the $x$ and $y$ components have finite values dependent on $\tau$. However, the time averages of the spin expectation values of the $x$ and the $y$ components over one cycle vanish. On the other hand, the spin expectation value of the $z$ component has a finite value and its time average over one cycle does not vanish. In Fig. 5.3c, the spin expectation value of the $z$ component changes periodically, with its cycle equal to one third of the cycle of the microscopic local rotation of the lattice. In addition, when the direction of the microscopic local rotation is reversed, the spin expectation values are reversed. Therefore, the microscopic local rotation works like an effective magnetic field, and the spin expectation values along the rotational axis are proportional to the angular frequency.

## 5.6 Discussion

We discuss how the microscopic local rotation affects electron spins. In equilibrium, the expectation values of the electron spins vanish without the spin-orbit interaction or magnetic field. In the system with the spin-orbit interaction, such as a Rashba system, the electron spins couple with the electron orbital motions due to the spin-orbit interaction, and the electronic bands are split by spins. Nevertheless, it is not trivial what kind of interactions enable coupling between the electron spin and the microscopic local rotation. We use our toy model to clarify this issue. In our toy model, when the Rashba parameter $\lambda_R$ is zero, the spin expectation value along the rotational axis is zero since the energy bands are spin degenerate. Next, we compare the two cases: a simple vibration and a rotation of atoms A and B. When the atoms A and B mutually vibrate, the displacement vector $\mathbf{u}$ can be rewritten as

$$\mathbf{u} = \mathbf{u}_B - \mathbf{u}_A = -u_+(\cos\Omega t, 0). \tag{5.80}$$

We show the sum of spin expectation values over the first Brillouin zone in Fig. 5.4. Then the sum of spin expectation values over the first Brillouin zone has a finite

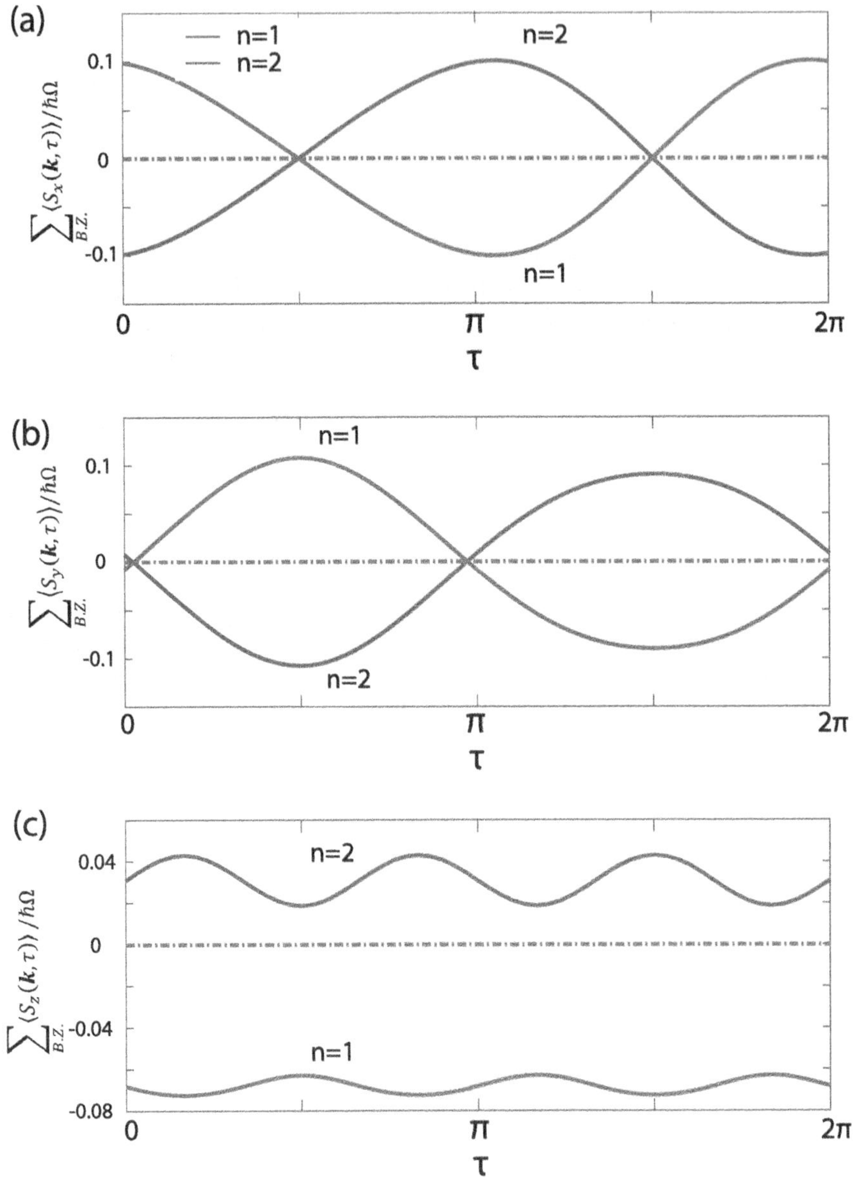


**Fig. 5.3** Sum of the spin expectation values over the first Brillouin zone in the system with the microscopic local rotation versus the rescaled time $\tau$ for **a** $x$, **b** $y$, and **c** $z$ components. We set the parameters as $\lambda_v = 0.2t_0$, $\lambda_R = 0.4t_0$, $\delta t' = 0.1t_0$, and $\delta\lambda_R = 0.1\lambda_R$. The dashed lines show $\sum_{B.Z.} S_i^{nn}(\mathbf{k}, \tau)/\hbar\Omega$, and the bold lines show $\sum_{B.Z.} \langle |S_i(\mathbf{k}, \tau)\rangle\rangle/\hbar\Omega$. Modified figure with permission from [11] Masato Hamada and Shuichi Murakami, "Conversion between electron spin and microscopic atomic rotation", Phys. Rev. Research 2. 023275, 2020, by the Author(s) licensed under CC BY 4.0

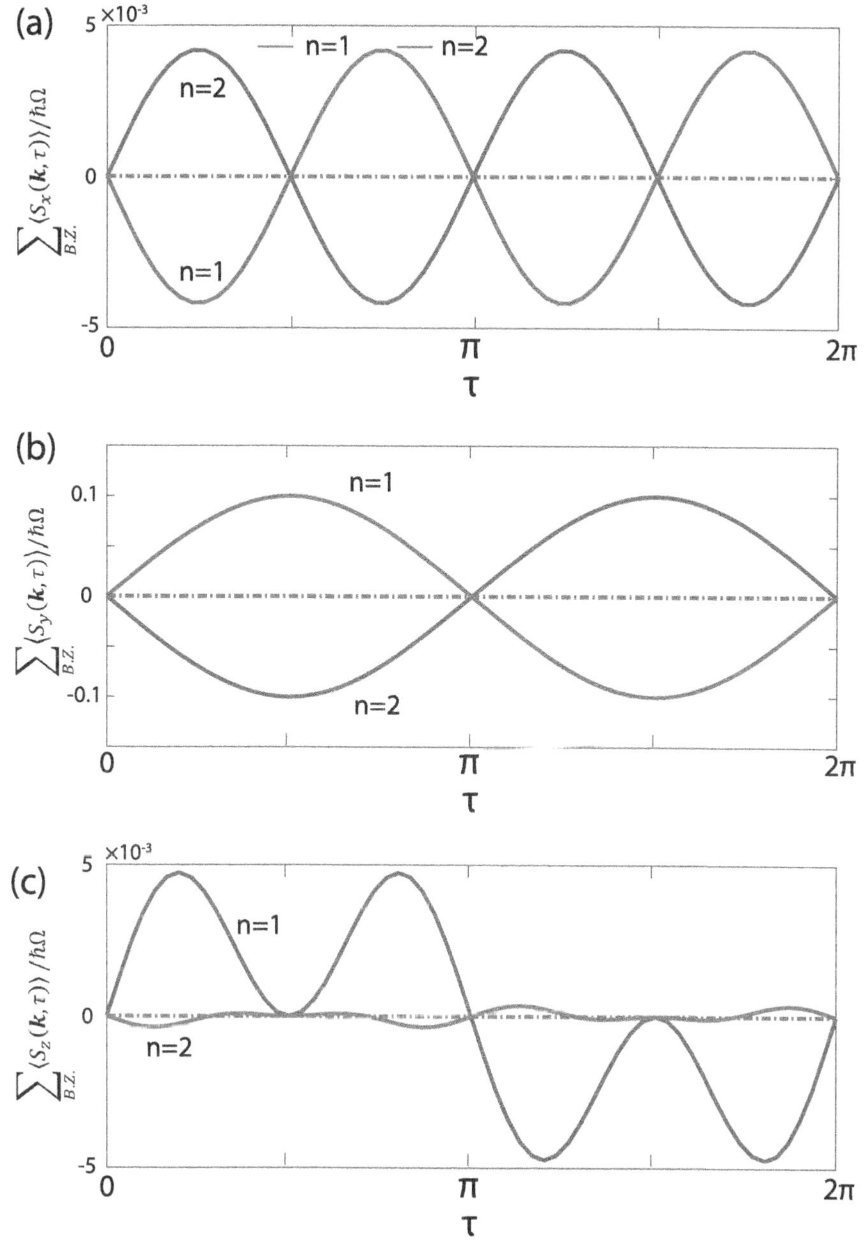


**Fig. 5.4** Sum of the spin expectation values over the first Brillouin zone in the system with the simple vibration versus the rescaled time $\tau$ for **a** $x$, **b** $y$, and **c** $z$ components. We set the parameters as $\lambda_v = 0.2t_0$, $\lambda_R = 0.4t_0$, $\delta t' = 0.1t_0$, and $\delta\lambda_R = 0.1\lambda_R$. The dashed lines show the $\sum_{B.Z.} S_i^{nn}(\mathbf{k}, \tau)/\hbar\Omega$, and the bold lines show the $\sum_{B.Z.}\langle|S_i(\mathbf{k}, \tau)\rangle/\hbar\Omega$. Modified figure with permission from [11] Masato Hamada and Shuichi Murakami, "Conversion between electron spin and microscopic atomic rotation", Phys. Rev. Research 2. 023275, 2020, by the Author(s) licensed under CC BY 4.0

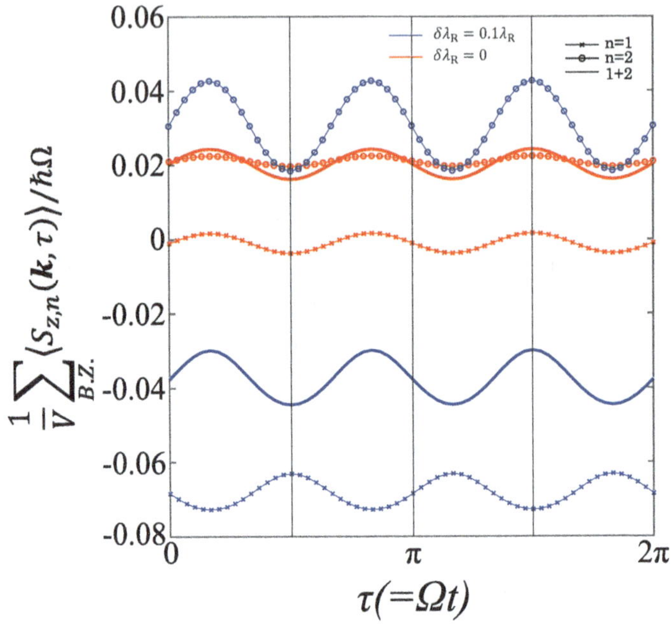


**Fig. 5.5** Sum of the spin expectation values of the $z$ component over the first Brillouin zone with and without the modulation of the Rashba term versus the rescaled time $\tau$. We set the parameters as $\lambda_v = 0.2t_0$, $\lambda_R = 0.4t_0$, $\delta t' = 0.1t_0$, $\delta\lambda_R = 0.1\lambda_R$ (blue line) and $\delta\lambda_R = 0$ (red line). The lines with crosses and circles show the lowest and second lowest band, respectively, and the bold lines show the sum of the lowest and the second lowest band. Modified figure with permission from [11] Masato Hamada and Shuichi Murakami, "Conversion between electron spin and microscopic atomic rotation", Phys. Rev. Research 2. 023275, 2020, by the Author(s) licensed under CC BY 4.0

value dependent on $\tau$ because the lattice vibration affects the electron spin via the spin-orbit interaction. However, their time average over one cycle becomes zero because of the time-reversal symmetry. At $\tau = 0, \pi, 2\pi$, the spin expectation values become zero because of the time-reversal symmetry. The spin expectation value is nonzero for the $y$ component at $\tau = \pi/2$ and $\tau = 3\pi/2$ since the system at $\tau = \pi/2$ and $\tau = 3\pi/2$ has TR $\times\, \sigma_y$ symmetry. Therefore, in the system with the simple vibration, the average of the spin expectation values become zero.

Next, we show the spin expectation value of the $z$ component without the modulation of the Rashba term in Fig. 5.5. In our model, the spin splitting is caused by the Rashba spin-orbit interaction. Therefore, it is considered that the modulation of the Rashba term by the microscopic local rotation mainly contributes to the generation of the spin expectation values along the rotational axis. However, in the system without the modulation of the Rashba term $\delta\lambda_R = 0$, the average of the spin expectation values of the $z$ component is nonzero as shown in Fig. 5.5. Hence, in order to couple the electron spin and the microscopic local rotation, the spin-orbit interaction, which connects the electron spin with orbital motions of electrons, is necessary.

## 5.7 Summary

In summary, we have theoretically found the coupling between the microscopic local rotation and the electron spins in the system with the spin-orbit interaction. The phonon angular momentum is a microscopic local rotation, and this can have a finite value at any **k** in a system without inversion symmetry. We consider a toy model on the two-dimensional honeycomb lattice with the Rashba spin-orbit interaction, and let the atoms A and B at the sublattice rotate around their equilibrium positions. Here, since the phonon energy $\hbar\Omega$ is typically much smaller than the electron energy, we calculate the spin expectation values using the second order adiabatic approximation. When the system does not have the spin-orbit interaction, the energy bands of the system are spin degenerate, and the microscopic local rotation cannot affect the electron spin. When the system has the spin-orbit interaction, which gives rise to spin splitting of energy bands, the spin expectation value become nonzero depending on time $\tau$. The time average of the spin expectation values perpendicular to the rotational axis over one cycle becomes zero. The time average of the spin expectation value along the rotational axis becomes nonzero and this is proportional to the angular velocity. On the other hand, in the system with a simple vibration of the atoms, the spin expectation values have nonzero values as a function of $\tau$, but their time averages are zero because of the time-reversal symmetry. Moreover, when the system does not have modulation of the spin-orbit interaction due to the microscopic local rotation, the spin expectation values become nonzero. Therefore, we have shown that the electron spins couple to the microscopic local rotation of atoms via the spin-orbit interaction, and the spin magnetization is proportional to the angular frequency of the atomic rotational motion.

## References


1. Einstein A, de Haas WJ (1915) Experimenteller nachweis der ampereschen molekularstrome 17:152–170
2. Barnett SJ (1915) Magnetization by rotation. Phys Rev 6:239–270
3. Matsuo M, Ieda J, Saitoh E, Maekawa S (2011) Effects of mechanical rotation on spin currents. Phys Rev Lett 106:076601
4. Matsuo M, Ieda J, Saitoh E, Maekawa S (2011) Spin-dependent inertial force and spin current in accelerating systems. Phys Rev B 84:104410
5. de Oliveira C, Tiomno J (1962) Representations of dirac equation in general relativity. Il Nuovo Cimento 24:672–687
6. Mashhoon B (1988) Neutron interferometry in a rotating frame of reference. Phys Rev Lett 61:2639–2642
7. Matsuo M, Ieda J, Harii K, Saitoh E, Maekawa S (2013) Mechanical generation of spin current by spin-rotation coupling. Phys Rev B 87:180402
8. Kobayashi D et al (2017) Spin current generation using a surface acoustic wave generated via spin-rotation coupling. Phys Rev Lett 119:077202
9. Hamada M, Yokoyama T, Murakami S (2015) Spin current generation and magnetic response in carbon nanotubes by the twisting phonon mode. Phys Rev B 92:060409
10. Takahashi R et al (2016) Spin hydrodynamic generation. Nat Phys 12:52

11. Hamada M, Murakami S (2020) Conversion between electron spin and microscopic atomic rotation. Phys Rev Res 2:023275. https://doi.org/10.1103/PhysRevResearch.2.023275
12. Zhang L, Niu Q (2015) Chiral phonons at high-symmetry points in monolayer hexagonal lattices. Phys Rev Lett 115:115502
13. Kane CL, Mele EJ (2005) $Z_2$ topological order and the quantum spin hall effect. Phys Rev Lett 95:146802
14. Haldane FDM (1988) Model for a quantum hall effect without landau levels: condensed-matter realization of the "parity anomaly". Phys Rev Lett 61:2015–2018
15. Sasaki K-I, Saito R (2008) Pseudospin and deformation-induced gauge field in graphene. Prog Theor Phys Suppl 176:253–278
16. Berry MV (1987) Quantum phase corrections from adiabatic iteration. Proc R Soc Lond A 414:31–46
17. Rigolin G, Ortiz G, Ponce VH (2008) Beyond the quantum adiabatic approximation: adiabatic perturbation theory. Phys Rev A 78:052508
18. Trifunovic L, Ono S, Watanabe H (2019) Geometric orbital magnetization in adiabatic processes. Phys Rev B 100:054408

# Chapter 6
# Conclusion

In this thesis, we theoretically predicted new phononic effects leading to generation of phonon angular momentum in non-magnetic and magnetic crystals. We also proposed experimental methods to measure the generated phonon angular momentum. Moreover, we showed coupling between the electron spin and the microscopic local rotation. Until recently, the phonon angular momentum has not been much studied. It is partly because in a system without inversion symmetry, the phonon angular momentum is an odd function of the wavevector **k**, and the total phonon angular momentum in equilibrium completely vanishes.

In Chap. 3, we found that the phonon angular momentum is generated by temperature gradient in a system without inversion symmetry. When temperature gradient is applied to the system without inversion symmetry, the phonon distribution is deviated from equilibrium, and the total phonon angular momentum becomes nonzero. This mechanism is analogous to the Edelstein effect in electronic systems. We call it the phonon thermal Edelstein effect. In order to realize this effect, the symmetry of this system should be sufficiently low. We estimated the phonon angular momentum in two types of crystals: a polar crystal, such as wurtzite gallium nitride (GaN), and a chiral crystal, such as tellurium (Te) and selenium (Se). We showed the phonon angular momentum generated by temperature gradient for wurtzite GaN, Te and Se using the first-principle calculation. Since the phonon angular momentum cannot be directly observed, we proposed conversion of the phonon angular momentum to a rigid-body rotation of the crystal and to a magnetization. When the crystal can rotate freely, the phonon angular momentum generated by the temperature gradient is converted to a rigid-body rotation of the crystal due to the conservation of angular momentum. This rigid-body rotation is sufficiently fast for experimental measurement when the size of the sample is of the order of micrometers. On the other hand, because nuclei have effective charges, the phonon angular momentum generated by temperature gradient induces magnetization. This magnetization due to the phonon angular momentum is quite small. We also evaluated temperature dependence in the



M. Hamada, *Theory of Generation and Conversion of Phonon Angular Momentum*, Springer Theses, https://doi.org/10.1007/978-981-33-4690-1_6

high and low temperature limit. The response tensor for the phonon thermal Edelstein effect is proportional to $T^3$ at low temperature, and this converges to a constant in the high-temperature limit. Therefore, this effect is expected to be larger at high temperature than at low temperature.

In Chap. 4, we showed the phonon angular momentum in systems without time-reversal and inversion symmetries, such as magnetic crystals. When the temperature gradient is applied to systems with neither of these symmetries, the phonon angular momentum consists of two terms: an equilibrium term and a term proportional to the temperature gradient. For the phonon thermal Edelstein effect in a magnetic crystal, the response tensor is determined by its magnetic point group. On the other hand, when both symmetries are broken and their product is conserved, the phonon angular momentum of each phonon mode becomes zero at any wave vector **k**, and the total phonon angular momentum becomes zero. In this case, the phonon thermal Edelstein effect does not occur because the deviation of the phonon distribution does not contribute to the phonon angular momentum. As another mechanism to generate the phonon angular momentum, the phonon angular momentum is generated by the electric field. This mechanism is analogous to the magnetoelectric effect in multiferroic materials. We call it the phonon rotoelectric effect. The response tensor for the magnetoelectric effect of phonons is determined by the magnetic point group and has the same form as the response tensor for the magnetoelectric effect in multiferroic systems. We also showed that the phonon angular momentum is generated by an electric field using our toy model. In the phonon rotoelectric effect, the modulation of the phonon angular momentum of each mode due to lattice deformation is important. Moreover, its temperature dependence is the same as the phonon angular momentum in a magnetic field. Therefore, the phonon angular momentum due to the phonon rotoelectric effect vanishes in the high-temperature limit, and the zero-point phonon angular momentum is dominant at low temperature.

In Chap. 5, we showed microscopic mechanism for coupling between the electron spin and the microscopic local rotation of atoms. In previous works in spintronics, it has been shown that the magnetization and spin current are generated by a mechanical rotation via the spin-rotation couping, which couples the electron spin and mechanical rotation. In this thesis, we have discussed how the magnetization is induced by the phonon angular momentum which is a microscopic local rotation of atoms. Since the phonon frequency is typically much smaller than that of the electron motion, we calculated the spin expectation values in the two-dimensional honeycomb lattice with the Rashba spin-orbit interaction and the microscopic local rotation of atoms A and B in the sublattice by using the second order adiabatic approximation. We showed that the time average of the spin expectation value along the rotational axis over one cycle is proportional to the angular velocity of rotational motion of atoms. In the system with a simple vibration of atoms, we showed the spin expectation values become nonzero as a function of time. However, their time averages vanish because of the time-reversal symmetry. Moreover, in the system without spin-orbit interaction, the electron bands have no spin splitting and the spin polarization is not generated by a microscopic local rotation. Therefore, we expect that the spin magnetization is induced by the microscopic local rotation in systems with the spin-orbit interaction.

As a future outlook, we have found methods to generate the phonon angular momentum in non-magnetic and magnetic crystals. We expect yet other new effects which couple the phonon angular momentum with other degree of freedom, such as the Barnett effect of phonons and the inverse phonon thermal Edelstein effect. When non-magnetic crystals are rotated, the Coriolis force acts on each atom in the crystal and the time-reversal symmetry is broken. Therefore, the phonon angular momentum is expected to appear. Moreover, when non-magnetic crystals without inversion symmetry rotate, the phonon angular momentum is expected to be generated by the Coriolis force and then heat current may flow depending on time. In multiferroic materials, when the electric field is applied, both the magnetization and the phonon angular momentum is expected. Because in multiferroic materials, which the nuclei have a effective charge, the magnetization of phonons is expected. Therefore, when the sample size is of the order of micrometers, it may be necessary to consider the magnetization from the phonon angular momentum in observation of the magnetization.

Moreover, in metals, the phonon angular momentum will be partially converted to electron spin via the spin-rotation coupling, which is similar to the spin current generation proposed for the surface acoustic waves in solids, and for the twiston modes in carbon nanotubes. Thus, because the microscopic local rotation couples to electron spin due to the spin-orbit interaction, spin magnetization and spin current may be generated by the phonon angular momentum in systems with the spin-orbit interaction. Further theoretical and experimental investigation of these couplings are promising for establishing physics of phonons in real materials and for applications in spintronics.

# Curriculum Vitae

## Education

- Postdoc, Department of Physics, Tokyo Institute of Technology, 4/19–3/20
- Ph.D., Department of Physics, Tokyo Institute of Technology, 4/16–2/19
- M.A., Department of Physics, Tokyo Institute of Technology, 4/14–3/16
- B.A., Department of Physics, Tokyo Institute of Technology, 4/10–3/14

## Research Experience

### Postdoc

Advisor: Prof. Shuichi Murakami, 4/19–3/20

- Theoretical study of conversion of phonon angular momentum

### Ph.D.

Advisor; Prof. Shuichi Murakami, 4/14–3/19

- Theoretical study of generation of phonon angular momentum

## Work Experience

- Postdoc, Department of Physics, Tokyo Institute of Technology, 4/19–3/20



M. Hamada, *Theory of Generation and Conversion of Phonon Angular Momentum*, Springer Theses, https://doi.org/10.1007/978-981-33-4690-1

The manufacturer's authorised representative in the EU is Springer Nature Customer Service Centre GmbH, Europaplatz 3, 69115 Heidelberg, Germany. If you have any concerns regarding our products, please contact ProductSafety@springernature.com

Printed and bound by CPI Group (UK) Ltd, Croydon, CR0 4YY
30/09/2026
02225306-0003